Basic Principles of Semiconductors

By
Irving M. Gottlieb, P.E.

Author of:
Test Procedures for Basic Electronics

An Imprint of
Howard W. Sams & Company
Indianapolis, Indiana

REVISED FIRST EDITION - 1995

PROMPT® Publications is an imprint of Howard W. Sams & Company, 2647 Waterfront Parkway East Drive, Suite 300, Indianapolis, IN 46214-2041.

This book was originally developed and published as *Fundamentals of Transistor Physics* by John F. Rider Publishing, Inc., New York, NY.

International Standard Book Number: 0-7906-1066-3

Additional Edits: Karen E. Mittelstadt
Additional Illustrations: Doug Cobb
Cover Design: Phil Velikan

Printed in the United States of America

9 8 7 6 5 4 3 2 1

contents

preface

It is common knowledge that the preponderance of electronic technology now centers around solid-state devices. These contraptions, in turn, make use of a class of materials known as semiconductors, in which silicon has long enjoyed a prominent role. It is an open secret too, that such silicon devices have largely displaced vacuum tubes at the forefront of technology. One might speculate that archeologists of a future time will suppose that the tube represented the more advanced technology because of its intricate and complex configuration, especially when compared to what the eye beholds in semiconductor devices. Of course the very opposite is actually the case, and it is easy to become immersed in very deep and formidable studies when probing the principles of semiconductors.

Fortunately, it is also true that a large area of practical implementations can be well served from an investigation of the basics underlying the art of semiconductors. This, indeed, is the motif of this entire book. Such a theme will be found useful to a wide spectrum of practitioners in the applied science of electronics. Hobbyists, technicians, and experimenters will benefit, as will professionals with expertise in other disciplines, but who have not yet acquainted themselves with the basic principles of semiconductors. If, for example, one's goal is to master the technology of integrated circuits, it would first be necessary to grasp the fundamental ideas outlined in this book.

The neophyte will probably be surprised to learn that semiconducting materials are not merely "bad conductors," but rather unique alloys with very unusual electrical properties. Multidiscipline investigations involving physics, chemistry, material science, and electrical engineering have been required to optimize these properties for practical use. And the end is not yet in sight — extended performance and hitherto unexploited responses continue to evolve. Thus, we have seen rectifying diodes mutate into light-emitting diodes, and then into semiconductor injection lasers. Accordingly, it is to be hoped that this book will not only serve the needs of mundane activities, but will inspire further studies of the voluminous technology of semiconductors.

The author acknowledges extraordinary fortune in having derived counsel and guidance from some of the very pioneers in the field who were directly responsible for the creation of the junction transistor and its subsequent development. Although the information thereby obtained dealt with the infancy of semiconductor technology, the basic principles remain valid for the understanding of later and more sophisticated devices. The following experts were at once kind and forbearing in translating high-level theories into useful principles suitable for the entry level practitioner:

1. Dr. William B. Shockley — One of the three 1956 Nobel Prize winners for the invention of the transistor.

2. Dr. Robert Noyce — The former vice president and director of research at Fairchild Semiconductor Laboratories, who went onto become president of Intel Corporation.

3. Dr. Dean Knapic — President of Knapic Electro-Physics Corporation which pioneered silicon semiconductor technology.

4. Drs. J.L. Moll and Joseph Petit — Both were associated with the Electrical Engineering Department of Stanford University where early semiconductor technology received considerable developmental impetus.

The author, of course, accepts full responsibility for his interpretations of the concepts gleaned from these notable pioneers.

February, 1995 Irving M. Gottlieb
Redwood City, California

1. atomic physics

BASIC PRINCIPLES

The transistor and vacuum tube possess in common the wonderful property of *proportionate amplification,* whereby weak control signals are reproduced faithfully in form, but at a higher power level (see Fig. 1-1). Our proficiency with the older device, the vacuum tube, was enhanced by a general comprehension of its basic operating principles. These included such phenomena as thermionic emission, space charge, electron acceleration, and various laws regarding the interaction of electrostatic and mobile charges. In similar fashion, we will be able to utilize the transistor most expediently if we first become acquainted with the important concepts underlying its operation.

THE PRACTICAL TRANSISTOR

End Result of Extensive Knowledge. It is advantageous to contemplate analogous situations between operating mechanisms in the two amplifying devices (see Fig. 1-2). Some of these analogies will be somewhat loose and general in nature, whereas others will suggest startling evidence of similar phenomena. Inasmuch as the transistor evolved from physical knowledge of advanced degree compared with that needed to produce the vacuum tube, it follows that our grasp of the relevant basics must be more extensive than

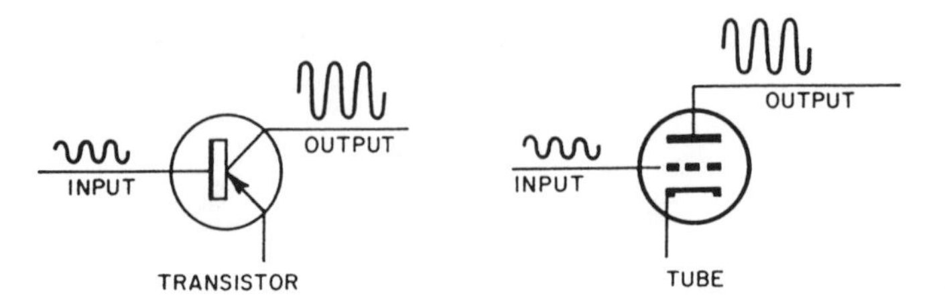

Fig. 1-1. The usefulness of both transistor and vacuum-tube stems from their ability to increase the power level of a weak signal applied to a control electrode.

that required for vacuum tubes. Accordingly, we begin our acquaintance with the transistor with a review and study of the very principles which had to be understood and applied to bring this marvelous device into creation.

We Must First Investigate Our Universe of Matter. Our exploration of the working principles of transistors must necessarily commence with consideration of the nature of matter. *Matter* is simply and broadly defined as the material substances from which our universe appears to be built up.

At the very outset, we should learn to cultivate appreciation of the essential differences between the "macroscopic" world of every-day experience and the submicroscopic world of matter building blocks. The significant aspect of the macroscopic world is that it is very much an *illusion* due to the restricted responses of our senses. The "realities" of our environment are, in general, fragmented and distorted pictures; it is up to scientific instruments to reveal closer views of the true situation. For example, a sheet

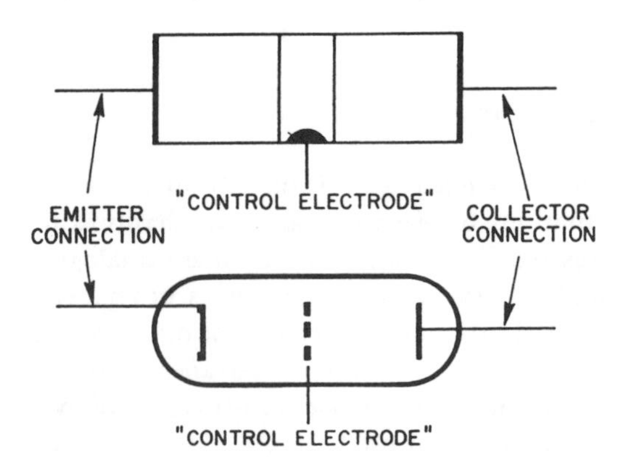

Fig. 1-2. Analogous electrode functions and operating principles prevail for transistors and vacuum tubes.

of steel feels "solid" to our sense of touch. Nevertheless, it is common knowledge that X rays can "see" through an apparently dense substance such as steel. This it can do only because the steel sheet is composed of constituent particles with vast spaces between them; in other words, contrary to information gathered from the tactile sense, the steel is *predominently space.* Although the relatively long wavelengths to which our eyes are responsive do not penetrate the steel, shorter wavelengths, such as X rays, find the steel quite transparent (see Fig. 1-3).

The events, impacts, and energy transformations observed in the macroscopic world are gross actions involving large aggregations of basic matter systems. Technology and science did not flourish until man sought knowledge in domains beyond those we casually, and oftimes insistantly, proclaim to be "reality."

Even Well-accepted Theories Fall Short of Ultimate Truth. In the submicroscopic world, we extend the response of our senses by means of scientific instruments. Through logic, we correlate

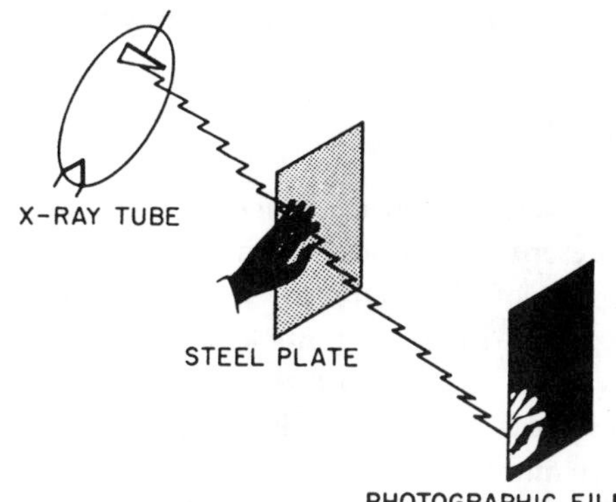

Fig. 1-3. The "solidness" of a steel plate is revealed by X-rays to be an illusion of our senses.

diverse phenomena to enable visualization of the arrangement and behavior of basic building blocks of the universe. By means of experiments, observations, and theoretical analysis we arrive at explanations which appear to tie together happenings of both submicroscopic and macroscopic worlds.

However, not only must we constantly make ourselves cognizant of the difference between the environments of elemental matter particles and of average or statistical effects of billions of such particles, but we must also balance our mental outlook in yet another way. We must retain awareness of the great difficulty in even *describing* the submicroscopic world in terms of words intended for the experiences related by our senses. Thus, when we investigate a fundamental-matter particle such as the electron, we tend intuitively to think of such an entity as a very minute globule of "substance." It is probably nearer the truth of the situation to state that the electron "particle" is not at all an almost unimaginable small speck of material, but rather a *wave packet*

Fig. 1-4. An atomic particle, such as the electron, is probably a unique manifestation of vibrant energy. However, it is useful to think of such an entity as a minute sphere of "substance."

of energy (see Fig. 1-4). However, such profound endeavor to get at the ultimate truth of creation will defeat the purpose at hand, for we would soon find it essential to reinforce our science with metaphysics, philosophy, and theology.

THE ATOM

Basic Building Block of our Matter Universe. The atom represents the fundamental system devised by nature for the building up of the material or matter universe. The configuration of the atom involves a nucleus, which can be likened to a sun, and electrons which orbit about the nucleus in the manner of planets in a solar system (see Fig. 1-5). The analogy to celestial bodies is even more striking when we consider that the electrons have an axial spinning motion as well as magnetic poles.

There are 92 different kinds of atoms, these corresponding to the 92 chemical elements. The gases, liquids, and solids of our environment are made up of these basic atoms and various combinations thereof. The simplest atom is that of hydrogen which contains a single electron in orbit about the nucleus. The nucleus of this atom behaves as an electrically charged particle and is approximately eighteen-hundred times "heavier" than the single orbital electron. Surprisingly, the distance between nucleus and orbital electron, proportionate to their sizes, is comparable to the distance between the sun and the earth!

Relationships of Electric Charges in the Atom. Not only the hydrogen nucleus but the nuclei of all atoms bear a positive electric charge. The orbital electrons, on the other hand, manifest the opposite electric charge, each electron representing a unit negative change (see Fig. 1-6). Each of the 92 elements involves a unique arrangement of orbital electrons, such that the sum of the electronic negative charges exactly balances the net positive charge of the nucleus. Thus, the element hydrogen is made up of atoms having a nuclear positive charge of one unit and a single orbital electron which bears one unit of negative charge. The helium atom has a nucleus which has a positive charge equal to

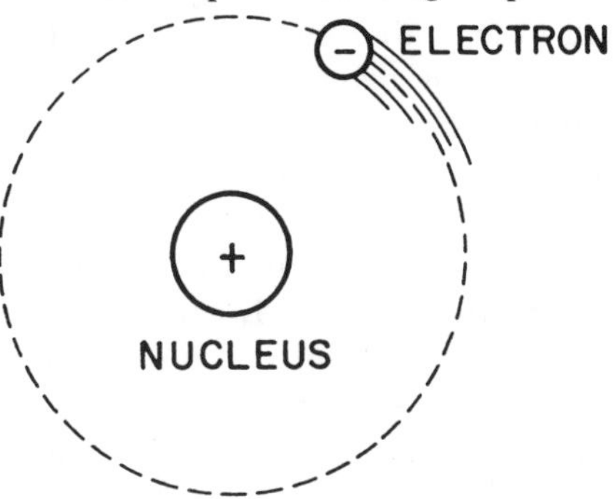

Fig. 1-5. The simple model of the hydrogen atom is depicted as a single electron in orbital motion about a nucleus, which bears a unit positive charge.

2 units. For this reason we find two orbital electrons in this atom. Again, the sodium atom consists of a nucleus with a net positive charge of 11, this being balanced by 11 orbital electrons. The heaviest and most complex atom, uranium, has a positive nuclear charge of 92 with a like number of electrons in orbital rings about the nucleus.

The Orbital Rings of the Atom. The simple model of the atom is depicted as a system of planetary electrons orbiting about a sun-like nucleus. The orbital motion of the electrons produces a centrifugal force just sufficient to prevent the electrons from "falling" into the nucleus due to the attractive force of opposite electric charge borne by the nucleus. The electrons describe their orbits within various concentric rings about the nucleus. The innermost orbital ring, the one closest to the nucleus, can contain only two planetary electrons. A second orbital ring may contain as many as eight electrons. The third ring can hold 18 electrons and the fourth ring, 32.

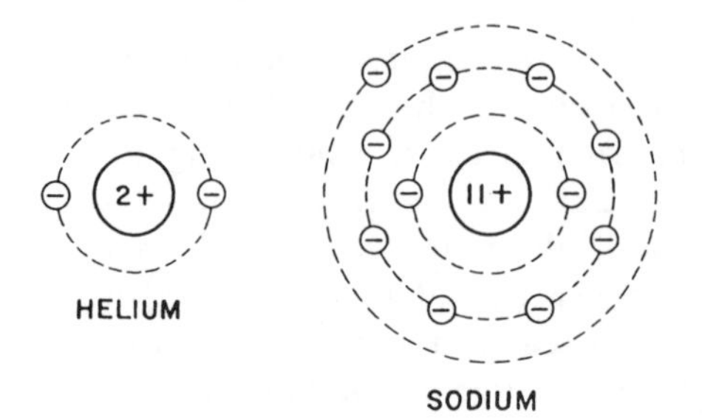

Fig. 1-6. The schematic representation of hydrogen is extended for more complex atoms.

A given orbital ring need not be filled up before electrons will occupy the next outermost ring, however. An important rule is that such planetary electrons cannot maintain orbits in spaces *between* the imaginary concentric rings surrounding the nucleus. When an atom is subjected to excitation by electrical forces, or from heat or light, electrons may be forced to jump from one orbit to another. This will be discussed in more detail.

At this time it is well to appreciate that a variety of phenomena is attributed to the behavior of the orbital electrons. This includes chemical reactions, magnetism, electricity, and emission of radiant energy such as light and X rays. The nucleus of the atom does not enter into interactions in any of these manifestations. The importance of the nucleus is that its positive charge determines the number of electrons contained in the orbital rings, which in turn governs the kind of atom built up (see Fig. 1-7).

THE NUCLEUS

Role of the Nucleus. It is expedient to think of the nucleus as a simple particle bearing a certain number of positive charge units Such a conception is workable because many observed and experienced phenomena are manifestations of the orbital electrons. However, it will not be amiss to briefly investigate some of the salient aspects of the nucleus. Although the nucleus itself is not directly involved in transistor action, it determines the orbital configurations of atoms, which in turn participate in the unique conduction phenomena of semiconductors.

Composition of the Nucleus. The nucleus is thought to be made up of at least several fundamental particle-like entities. Only those will be mentioned here which suffice to convey a simplified picture of the nucleus. Of major importance, from this point of view, is the proton. The proton is a fundamental entity, which bears a unit positive charge. The proton is approximately eighteen-hundred times as heavy as the electron.

Another particle, or particle system, residing within the nuclear domain is the neutron. The neutron apparently consists of a proton closely associated with an electron. Such a combination bears a net electric charge of *zero* and therefore contributes to the mass of the nucleus but *not* to its charge. This gives rise to the so-called *isotopes.*

Isotopes are members of the same chemical element which differ in atomic weight but not in chemical properties. For example, the oxygen atom has a nucleus with a net positive charge

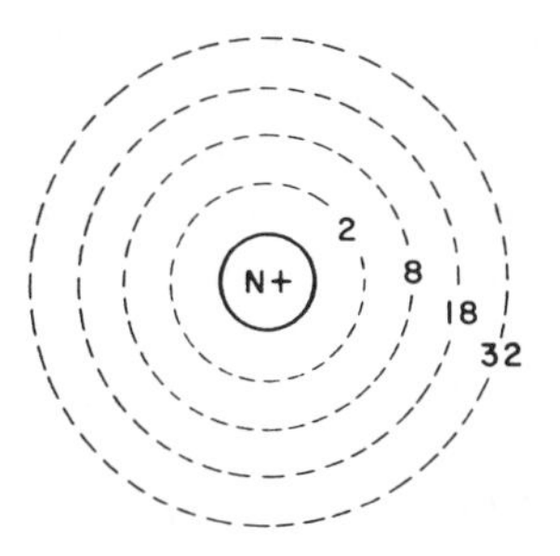

Fig. 1-7. Only a certain maximum number of electrons may be contained in the various orbital rings.

of 8. This charge condition permits an orbital electronic arrangement consisting of eight electrons. Two of these electrons are in the inner of two orbital rings, the remaining six are in the outer ring.

For oxygen there exist three kinds of nuclei, all having a *net* positive charge of *eight* and all supporting the *same* orbital electronic arrangement. These three nuclear varieties differ only in the number of *neutrons* they contain. One isotope contains eight neutrons; another, nine; and the third, ten. In a sense, the nucleus is "shielded" from chemical and electrical participation by the orbital electrons (see Fig. 1-8).

CHEMICAL ACTION

It is not at all easy to visualize why different electronic configurations should give us the characteristics associated with the various elements. Although explanations are well grounded in experiments and mathematics, it is exciting to the imagination to contemplate that the number and arrangement of revolving electrons within atoms produce the diverse properties displayed by the substances around us. It is indeed a startling fact that chemical

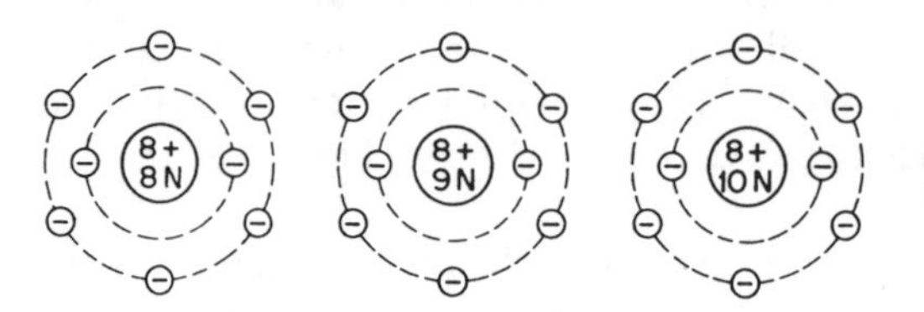

Fig. 1-8. Atoms representing the three isotopes of oxygen.

reactions are interactions of *outer* ring electrons, such electrons being displaced, interchanged, or shared when atoms come together. Thus, chemical action is in essence, an electronic phenomena. This is convincingly demonstrated by the electromotive cell or battery in which materials are so oriented that the interatomic exchange of electrons circulates through an external circuit. In simpler cases, we mix two chemicals and observe the generation of heat. We may safely postulate that this heat is produced by the movement of electrons, that is, the flow of electricity, within the reagent itself.

Stable States of the Atom. The question must naturally arise as to why atoms should enter into combination with one another when an apparently stable charge condition exists due to equalization of nuclear and orbital electronic charge. It is true that this condition confers sufficient stability to enable the atom to exist as such. However, in general, mere equilibrium of nuclear and electronic charges does not represent the lowest energy level possible for the atom.

It is axiomatic that atoms tend to assume a configuration representing the lowest energy needed to maintain a given nuclear charge in association with a group of orbital electrons. Although the atoms *do* indeed seek an arrangement permitting equilibrium of nuclear and electronic charges, they *also* seek another condition of equilibrium which is not obvious without resort to elaborate mathematics. This is an arrangement of *eight* electrons in the outermost orbit.

(Hydrogen and helium atoms are exceptions because their orbital electrons are contained in the first orbital ring. The maximum number of electrons allowable for the first ring is two. It so happens that, for *these* atoms, possession of *two* orbital electrons corresponds to maximum stability. The helium atom, for example, with its two orbital electrons is already satisfied and has no need to unite with other atoms. Consequently helium is chemically inert, forming no compounds with other elements.)

The Mechanism of Chemical Combination. Chemical combination occurs between atoms which can derive *mutual benefit* from the association, this benefit being in the nature of a lower energy state. For example, common table salt is the union of sodium and chlorine atoms. The sodium atom has a single electron in the third and outermost orbital ring from the nucleus. If it could shed this lone electron, a stabler state, one representing a lower energy level, would be attained with eight electrons in the second orbital ring. Because of the attractive force of the positive nucleus, this electron can be shed only under the influence of a stronger force. In similar fashion, the chlorine atom lacks an electron needed to fill its outermost orbital ring to eight electrons. Due to the nuclear positive charge of 17 units, the additional electron, that is, the 18th in the atom, can be acquired only under the influence of a stronger force. When a sodium atom and a chlorine atom come together, the "extra" electron from the sodium atom moves into the outermost orbit of the chlorine atom. Both atoms have supplied strong forces to overcome one another's charge. Both atoms are now "satisfied" with *eight* outer orbit electrons (see Fig. 1-9).

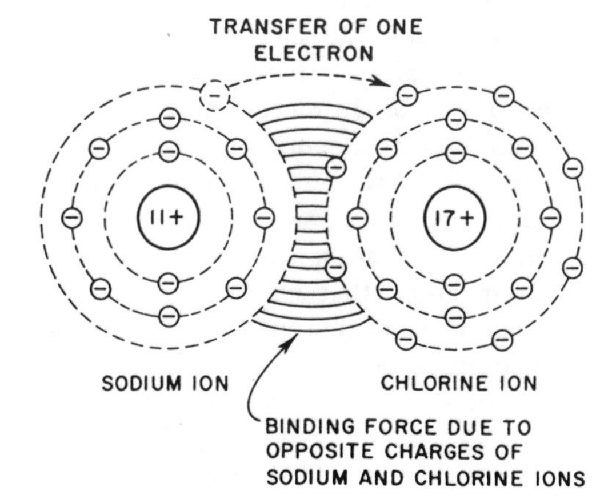

Fig. 1-9. The electrovalent type of association provides mutual stability for both combining atoms. After transfer of one or more electrons from one atom to the other, both atoms satisfy their tendency to maintain outer orbits with eight electrons.

We see that the altered sodium atom now has a net positive charge of 1 unit, whereas the altered chlorine atom now has a net negative charge of 1 negative unit. These charge conditions constitute the force of chemical combination. As long as nothing is done to the resultant compound, sodium chloride, each atom satisfies its charge "hunger" on the other atom, thereby producing a condition of *charge stability*.

We note that atoms in such chemical combination are not strictly the same atoms of the constituent elements. This accounts for the fact that compounds have radically different properties than their constituent elements.

Instances will be described in which atoms combine but do not transfer or receive orbital electrons from one another. In such cases (diatomic atoms) the net substance retains many of the properties of the constituent elements. The entity formed through association of two or more elemental atoms is known as the *molecule*. Molecules are the smallest portion of a substance which retain the properties of the substance.

Ionization. We have noted that in one type of chemical reaction the atoms undergo modification when they combine. The modification is such that the over-all molecule is electrically neutral. This is because the positive and negative charges of the modified atoms cancel. The modified atoms are called *ions*. The chemical union thereby produced is said to be maintained by electrovalent or ionic forces. Ions are of great significance because they can be caused to exist *apart* from the molecule itself. For example, the sodium chloride union can be split in such a way that sodium ions and chlorine ions are produced. We see that ions are essentially *electrically charged atoms*. As such, they may serve as *carriers* of the electric current, the flow of electricity being a manifestation of moving charges.

Inasmuch as the ion has a *different* orbital electron arrangement from the atom, we do not expect the ion to display the properties of the atom from which it was derived. For example the sodium atom, that is, the element sodium, is poisonous. Yet, we do not become poisoned from table salt because the sodium is in its ionic state.

One of the ways in which ionization is produced is by dissolving a compound in water (see Fig. 1-10). Water as a solvent, provides a special condition whereby the electrical force of attraction between ions of a molecule is greatly *weakened.* Water has a very *high* dielectric content; that is, a given cross-sectional area of water can support many lines of electric force. This diminishes the *intensity* of the force between regions of electric charge.

An analogous situation which may be suggested is magnetism. The force existing between the poles of a horseshoe magnet is greatly decreased when a substance offering ready passage to magnetic lines of force, such as a block of iron, is placed between the poles.

The water may be said to provide a medium in which the electrical forces of the ions are weakened to the point that ions with unlike charges are no longer impelled to remain in molecular combination. The disassociated ions are now available as *mobile carriers* of the electric current.

A demonstration of this phenomena is readily accomplished by immersing two chemically inert electrodes, say of carbon, in

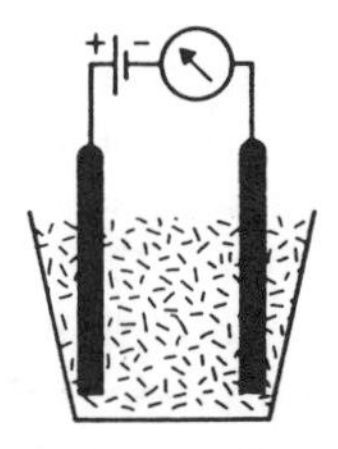

SODIUM CHLORIDE CRYSTALS

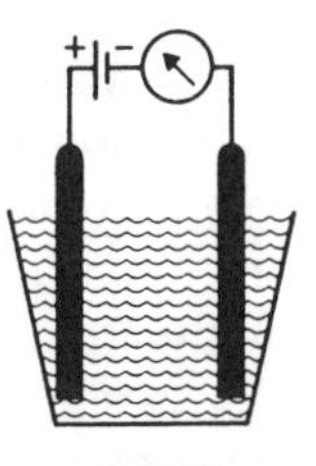

WATER

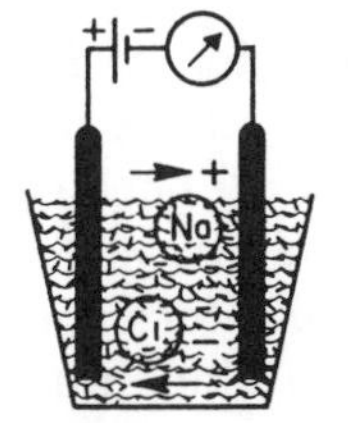

SODIUM CHLORIDE SOLUTION

Fig. 1-10. Neither sodium chloride nor water provides charge carriers for the conduction of electricity. In the solution, however, disassociated sodium and chlorine ions move to the electrodes, thereby enabling the flow of the electric current.

a solution of water and table salt. A battery and a meter connected to the external circuit reveal the solution to be a conductor of electricity. The experiment is made more convincing by also showing that ordinary table salt is a nonconductor, as is pure water. The solution of water and salt conducts electricity because, under influence of the applied electric field, the negatively charged chlorine ions are attracted to the positive electrode, and the positively charged sodium ions are attracted to the negative pole. The mutual transition of the two ions constitutes the flow of electricity within the solution. When the negative chlorine ion reaches the positive pole, it sheds its surplus electron and becomes an electrically neutral chlorine atom. When the positive sodium ion reaches the negative pole, it acquires an electron to fill the "hole" in its outer orbit. It then becomes an electrically neutral sodium atom. (The chlorine atoms escape into the atmosphere as chlorine gas. The sodium atoms undergo further chemical action in the water, forming sodium hydroxide and hydrogen.)

Dynamic Concept of Ion Production and Recombination. It is postulated that the ions in a solution are not static but rather are in a *constant cycle* of formation and extinction (see Fig. 1-11).

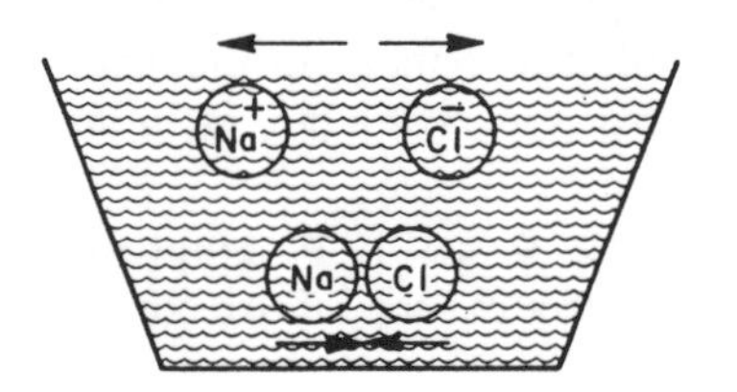

Fig. 1-11. Free ions continuously disassociate and reassociate.

Formation of ions is caused by weakening of the force of electrovalent charge by the water. Extinction is caused by recombination of ions. The tendency towards recombination always exists, due to the attractive force between oppositely charged ions. The rate of ion formation and recombination sets up a condition of equilibrium which governs the extent to which ionization exists in the solution. Thus, the charge-carrying ions can be said to have *certain lifetime.*

The Covalent Bond. Chemical union is not always attended by *transfer* of electrons from one atom to the other. Atomic associations are also made on a sharing basis wherein one or more outer orbit electron becomes the *common property* of two atoms. Whereas the ionic binding force of electrovalence involves two *unlike* atoms, *covalent* binding force can unite like *or* unlike atoms (see Fig. 1-12). The hydrogen molecule, for example, consists of a pair of hydrogen atoms thusly associated. Later, we shall find the covalent bond a significant feature of the germanium and silicon used in transistors.

Basic Action Tendencies of Atoms. The different ways in which interatomic associations are established is apt to create confusion. It is well to appreciate that atoms form combinations in order to equalize imbalances in force fields and to bring the atomic system to a state requiring less internal energy (see Fig. 1-13).

To drive home these facts, we note that certain atoms, those corresponding to the *inert* elements—helium, neon, argon, krypton, xenon, and radon,—are already satisfied and do not participate in chemical reactions. These atoms, with the exception of helium, all have *eight* electrons in their outermost orbit. (As previously mentioned, eight electrons in all outer orbits, except the closest permissible one to the nucleus, represents a condition of stability. For the closest permissible orbit the highest stability results from *two* electrons.) When this condition *coexists* with the balance of nuclear and orbital electronic charges, maximum atomic stability

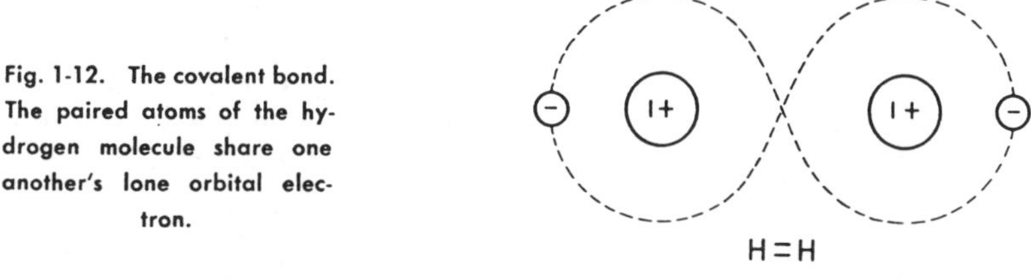

Fig. 1-12. The covalent bond. The paired atoms of the hydrogen molecule share one another's lone orbital electron.

results. This, we indeed recognize, is the case with the inert elements. Not only do atoms of these elements abstain from chemical combinations with other elements, but other phenomena involving outer orbit electrons are likewise absent.

Thus, there is no disassociation into ions without excitation from an external source of energy and no covalent association. Paradoxically, despite their nonentry into reactions, the inert elements have high academic significance. This is because all other elements are chemically active due to their tendency to assume the electronic arrangement of the *closest* inert element. This is the basic motivation; despite the fact that the stability of the nearest inert element is never actually reached, all chemical reactions seek the ultimate satisfaction in energy balance inherent in the inert elements.

Now, let us reexamine previously discussed reactions in this light. For example, consider the covalent bond of the hydrogen molecule. Here, the individual hydrogen atoms share one another's single orbital electron. This sharing *approximates* the conditions of the nearest inert element, helium, wherein *two* electrons are contained in the single orbit. However, the covalent bond satisfies only the tendency to effectively duplicate the arrangement of

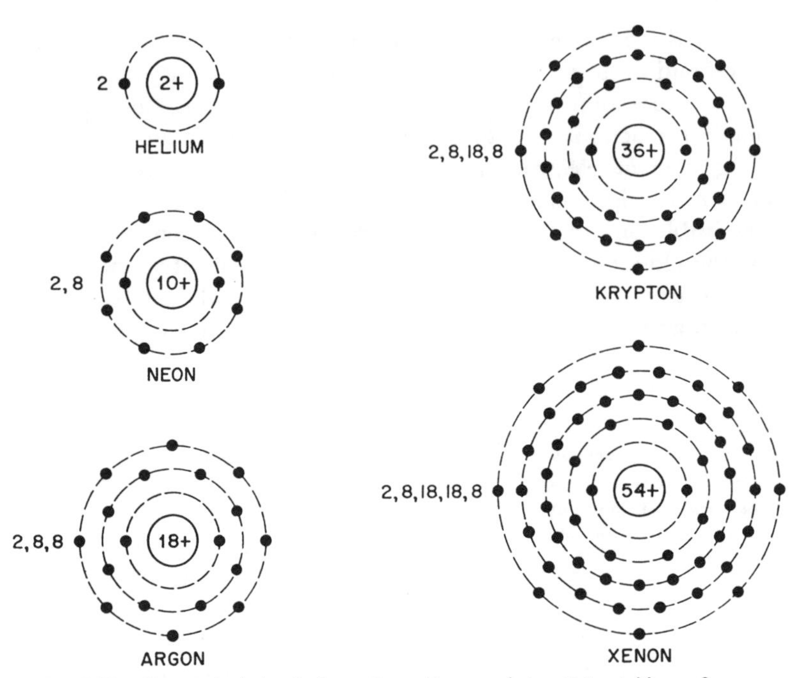

Fig. 1-13. The goal of chemical reactions. Atoms seek to attain stable configurations of the inert elements.

orbital electrons in the helium atom. Because there are now effectvely *two* orbital electrons per hydrogen atom, the hydrogen molecule manifests a net negative charge. Therefore, the hydrogen molecule itself seeks associations with other atoms in order to attain *charge equilibrium.* We observe that the pair of atoms making up the hydrogen molecule cannot be inert, as is helium, because the nucleus of the hydrogen atom bears only a single positive charge. The helium nucleus has a positive charge of two units, thereby balancing the negative charge of its two orbital electrons.

In our discussion of the chemical union involved in producing sodium chloride, we noted that the sodium atom released its lone outer-orbit electron which moved to the outer orbit of the chlorine atom, filling the electron deficit, or "hole," in that electronic arrangement. We are interested now to see that this transaction results in the sodium atom partaking of the electronic configuration of the inert element *neon.* At the same time, the chlorine atom has been modified to have an orbital electron arrangement identical to the inert element *argon*. Nevertheless, the elements neon and argon are nonexistant in the compound sodium chloride. The modified sodium atom, now the sodium ion, differs from the neon atom in the matter of the positive nuclear charge. The same is true of the chlorine ion; its resemblance to the helium atom ceases when nuclear charges are compared.

Reaction Between Energy and Atomic Systems. We have already noted that the particle nature of atomic entities, such as electrons, is but one aspect of their existence. Although we prefer to deal with the electron much as we would a billiard ball and describe its behavior in terms of mass, impact, and kinetic energy, the wave-motion characteristic of radiant energy is also associated with the electron. Indeed, a more rigorous treatment of electronic phenomena within the atom dispenses largely with treatment of particle behavior but explains its existence and manifestations in terms of the mathematics of wave motion.

Conversely it is also true that radiant energy such as light produces effects related to noncontinuous causation; it is as if the light energy propagates in tiny successive bundles of energy, quite suggestive of a train of discreet particles.

Thus, what we "intuitively" think of as *two* domains of nature, *matter* and *energy*, are not so different after all. Perhaps, it is a carryover from our illusionary macroscopic world of everyday experience that causes us to insist that matter and energy are qualitatively different. Not only is there an exact equivalence between "matter and energy," but diverse physical phenomena can be successfully interpreted in terms of *either* particle or wave analysis. This being the case, we should not be surprised to find many interactions between radiant energy and atomic systems, for they are essentially of the same ingredient.

Most chemical reactions are speeded up by the addition of heat. Here, thermal energy has acted on the orbital electrons, moving some in larger orbits, thereby weakening the nuclear binding force. When the cathode of any vacuum tube is heated, both particle-like and wave-like disturbances ensue; that is, electrons, and light are emitted. The light is due to thermal excitation of *inner* orbit electrons, whereas the electron emission results from excitation of *outer* orbit electrons.

Here we observe that the thermal reaction with similar entities at different energy levels gives rise to two emissions. From the similar causation, we are led to suspect that the effects are closely related. In the vacuum tube, the electron stream impinging on the anode produces heat. It is as though the energy has reverted to its earlier form. If the anode voltage of the vacuum tube is sufficiently high, the innermost electron orbit of the anode metal is violently excited, and X rays are produced. These high-frequency

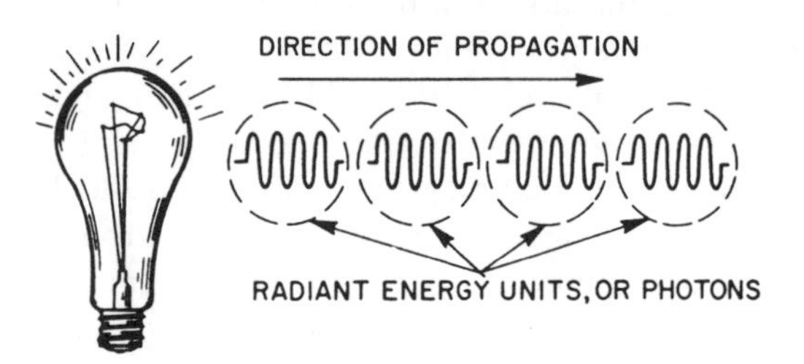

Fig. 1-14. Heat and light exhibit a dual nature, manifesting both corpuscular and wave behavior.

waves display a particle-like property when they encounter orbital electrons of gas atoms, for the reaction is in the nature of an *impact* wherein electrons are ejected from the gas atoms, producing ionization.

We have considered sufficient phenomena to appreciate the interplay and overlap between so-called *particles* and *waves*. Having decided for the sake of simplicity to deal with atoms as particle systems, it will facilitate our discussions of energy interactions with atoms if we emphasize the particle-like properties of radiant energy. For this purpose, we think of radiant energy as transmitted in discrete packages called *photons*. In this way we will be dealing essentially with a particle-to-particle impact phenomenon which lends itself to nonmathematical visualization more readily than analysis based on wave-to-wave or wave-to-particle reactions (see Fig. 1-14).

Large Aggregates of Matter. A gas consists of tremendous numbers of atoms, atomic pairs, or molecules existing more or less independently. Liquids and solids require closer associations between such entities.

Let us investigate some aspects of the structure of sodium chloride crystals. Such crystals are gross chunks of matter which owe their existence to a repetitive pattern of ionic arrangement. This pattern or orientation of ions is called the *crystal lattice.*

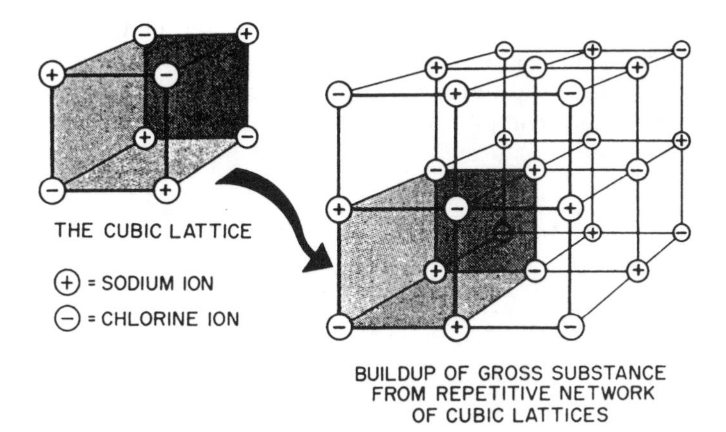

Fig. 1-15. Crystalline structure of sodium chloride.

For sodium chloride the crystal lattice is a symmetrical cube. Since the crystal is constructed of such ionic cubes, each sodium ion is surrounded by six chlorine ions; likewise, each chlorine ion is surrounded by six sodium ions. In this type of matter buildup, it is more meaningful to consider the crystal lattice rather than the molecule as the basic building block (see Fig. 1-15).

Germanium and silicon crystals are used in preparing the material from which transistors are made. In pure form, such crystals consist of like atoms arranged in repetitive patterns of the cubic lattice and maintained by covalent binding force. Crystal lattices of various substances may be constructed from ions, atoms, or molecules and may involve a single element, or elements, in chemical union. The essential ingredient always involves a repetitive arrangement of systematically oriented particles. Most solid substances are at least capable of assuming crystalline structure. When crystalline structure is lacking, the substance is said to be *amorphous*. In such a case, the structure is generally a haphazard one of cohering molecules. Molecules as well as all other matter particles or aggregates have an affinity for one another. This force is in the nature of a gravitational field and is relatively weak compared to the electrical forces within the atom. In a gas, the molecular forces are negligible in comparison with the influence of thermal energy which sets the individual gas molecules in violent motion. If the gas molecules are brought together by pressure and if heat is removed by cooling, the molecular forces of attraction do assert themselves, and the gas is converted into the liquid state. Additional cooling further removes the disturbing effect of thermal energy with the result that the substance attains its solid state.

2. electrical conduction

THE ESSENTIAL MECHANISM OF CONDUCTION

Our investigation of basic physical and chemical principles underlying atomic behavior now leads us into discussion of electrical conduction, perhaps the most important atomic phenomenon involved in transistor action. We have already touched upon some aspects of electrical conduction, having found that the electric current is produced by moving charge carriers. Let us pursue further the study of electrical conduction in several substances. First we shall account for the most common type of conduction in electrical practice, the flow of current through metallic wires or objects; then we shall concern ourselves with liquids and gases.

CURRENT FLOW

In Metals. A metal such as copper is an elemental substance, being composed of copper atoms. The formation of the gross substance, copper, involves relatively tight packing of these atoms. The atoms are thought to be so close together that the outer electronic orbits overlap. Such electrons are not uniquely bound to any one particular atom; rather these electrons exist in the material somewhat as a "fog" surrounding the atoms. Consequently, these electrons are quite free to move from atom to atom. At ordinary temperatures, these electrons are imparted random motions by the action of thermal energy. This erratic displacement of free electrons produces pulses of electric current which may be audibly or visibly detected with the aid of a high-gain amplifier. This so-called "noise" actually embraces a very wide frequency spectrum.

If we subject the free electrons to the pressure of an electric field by connecting a battery to the wire, an orderly procession of electrons is superimposed upon the thermally induced random motion. Electrons are then supplied by the negative pole of the battery and "collected" at the opposite pole. Within the wire, electrons are impelled to move from one atom to another. Other than thermally induced vibration, the bulk of the atom is not free to move. It must not be thought that any particular electron injected into the negative end of the wire makes a speedy transition to the positive end. Rather, what we speak of as a *flow of electrons* is more precisely a *relaying* process wherein the impetus received at the negative end of the wire is communicated with great speed to an electron at the positive end of the wire. The latter electron is ejected from the wire and collected at the positive pole of the battery.

Thus, while it is true that electricity traverses such a conductor with a speed close to that of light, the motion of individual electrons from one end of the wire to the other is more in the nature of a slow drift (see Fig. 2-1). The basic mechanism involved may be demonstrated by arranging a row of marbles on a table. An impact conveyed to an end marble is quickly transmitted to the marble at the opposite end which is "expelled" from the row. The impacting "shooter" marble did not have to traverse the length of the row in order to eject the marble at the opposite end. Accompanying the ejection of the final marble was a "drift" of all marbles in the same direction toward which the final marble was ejected.

In Liquids. We have, in the previous chapter, found that electrical conduction in certain liquids is a consequence of the ionization process which provides mobile charge carriers. Generally, the current is made up of two components: one provided by the positive ions, and one provided by the negative ions. The movement of positive ions toward the negative pole and the negative ions toward the positive pole constitutes a migration of gross matter, a phenomenon of which practical use is made in electroplating. As in the case of metallic conduction, the flow of charge carriers is very slow compared with the so-called speed of electricity, for it is the propagation of the impetus from ion to ion which manifests itself as the electric current. Liquids, such as alcohol, oil, or a sugar solution do not contain free ions. Such liquids are insulators since they have no mobile charges to conduct the electric current.

In Gases. In gases the atoms lead existences relatively independent from one another, and each atomic nucleus exerts strong binding forces on its orbital electrons. Therefore, a gas in its normal state does not carry an electrical current.

A gas can, however, be forcibly ionized, whereupon mobile charges are provided to support electrical conduction. One method of producing ionization is simply to subject the gas to a strong electric field. Under the pressure of the electric field, outer orbit

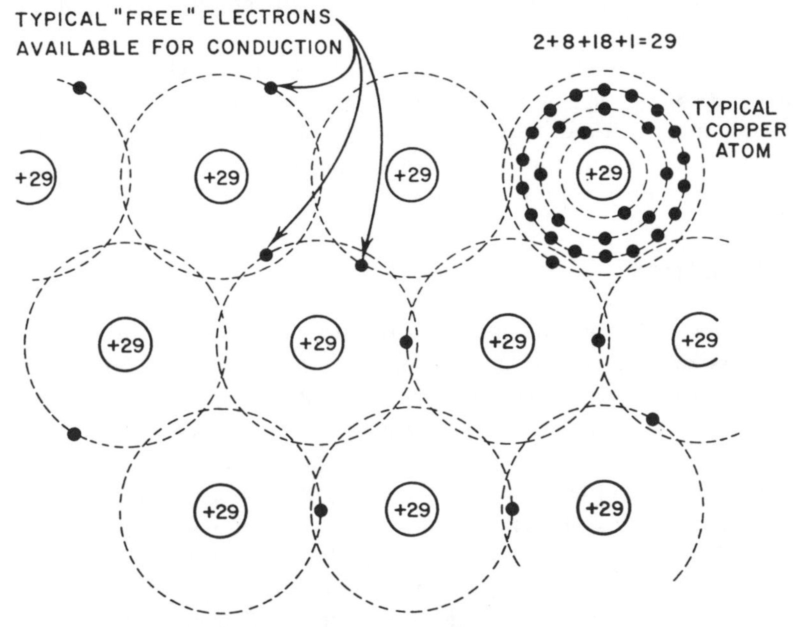

Fig. 2-1. Copper atoms are densely packed. The outer orbit electrons are free to move from atom to atom because these orbits overlap one another.

electrons are torn from the atoms and attracted to the positive pole of the voltage source. The atoms are in this manner converted to positive ions, which then move toward the negative pole.

Thus an electric current consisting of two component movements of charge carriers exists in the gas. The electrons and positive ions move in opposite directions, but their effects are additive insofar as total electric current is concerned.

Gases, even those classified as inert, are susceptible to ionization by impinging thermal photons as well as photons produced by higher-frequency radiant energy such as radio waves, light, and X rays. When subjected to strong excitation from such energy sources, outer-orbit electrons are dislodged from the binding force of gas atoms. Such electrons may absorb sufficient energy from the radiation to subsequently dislodge one or more additional electrons from other atoms. The dislodged electrons as well as the resultant positive ions may then be directed to polarized electrodes. The voltage impressed across the electrodes may be considerably lower than that required for ionization in the absence of excitation by radiant energy.

We see that, no matter how produced, mobile charge carriers are necessary for conduction. It is also apparent that, whereas metals and ionized liquids (electrolytes) are ready to provide electrical conduction, energy must be added to a gas before mobile charge carriers are produced to carry current (see Fig. 2-2).

INSULATORS

Substances which normally do not provide mobile charge carriers do not permit passage of the electric current. Such substances are known as *insulators*. In our study of transistor action it will be important to appreciate that even insulators generally become conductive when subjected to appropriate energy bombardment. For example, as we pointed out, gases are insulators prior to penetration by photons, or to the influence of a strong electrical field. Also, solid insulators suffer degradation from the application of heat, and as the temperature approaches the melting point, relatively good conduction generally occurs. The germanium and silicon crystalline pellets used in transistors depend upon room-temperature-induced thermal photons to help free atom-bound electrons.

Our discussion of atomic structure was premised upon atoms in their so-called *ground state* wherein each orbital ring was of minimum allowable diameter. Such, indeed is the state of many types of atoms under normal conditions because energy interactions from thermal photons at room temperature are dwarfed by the strong nuclear binding forces. However, atoms subjected to intense excitation must somehow absorb the incipient energy.

The easiest way this can be done is by the orbital electrons jumping into larger diameter orbits. It still holds true that orbit expansion occurs in "jumps," only certain definite orbits being available to the excited electrons. Of particular importance, when

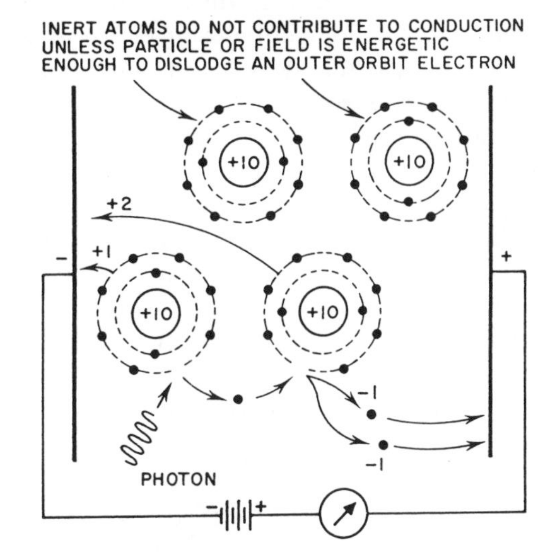

Fig. 2-2. Conduction in neon gas.

the outer orbital electrons are driven farther from the nucleus in their planetery excursions, the nucleus relinquishes a considerable amount of its binding force. Such electrons are readily dislodged from their atomic associations by moderate electric fields. Once freed, they are available as mobile charge carriers for conduction of the electric current.

GERMANIUM

Conductive Properties. We will discuss the conduction of electricity within germanium specifically, but it should be understood that conduction phenomena are essentially the same in silicon. Later, we shall explore some of the properties of germanium and silicon which account for differences in characteristics in transistors made of the two substances.

The germanium used in transistors consists of a small wafer sliced from a large crystal. The germanium crystal is a repetitive pattern of atoms arranged in symmetrical cubes. Such a building block, or crystal lattice, is similar to that already described for sodium chloride, except that the atoms of the lattice are all of one kind. The force which constrains the atoms in their lattice positions is derived from covalent bonds in which each germanium atom shares its four outer-orbit electrons with four equally distant neighbors. In this way, each atomic nucleus has effectively a planetary arrangement of electrons with eight outer orbit electrons.

Such an arrangement satisfies the tendency of the germanium atoms to emulate the nearest inert element, which in this case is krypton. In this electron-sharing association, the balancing of positive and negative charges within each atom is not disturbed. Each atomic nucleus still retains its positive charge of 32 units as in the case of an isolated germanium atom. Each atomic nucleus balances 32 orbital electrons as would be the case in an isolated germanium atom. Nevertheless, each germanium atom of the crystalline substance shares *four* extra electrons *from four* neigh-

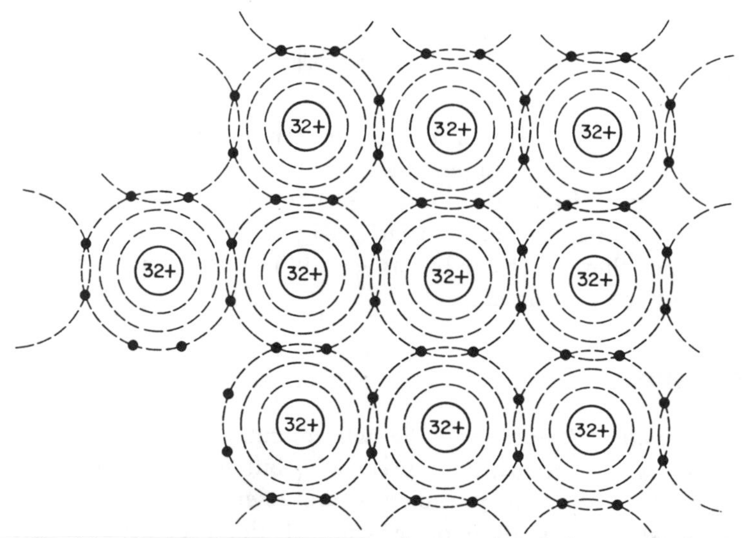

Fig. 2-3. Covalent sharing of electrons in germanium. Each atom satisfies its desire for eight outer-orbit electrons.

bors, simultaneously sharing its own four outer-orbit electrons *with* four neighbors (see Fig. 2-3).

We recall that the binding force in the sodium chloride crystal was ionic, the crystal lattice being composed of ions. In contrast, the binding force in the germanium crystal is of the covalent variety, and the crystal lattice is composed of *uncharged* atoms. Inasmuch as germanium appears to have no charge carriers, mobile or otherwise, we might expect this substance to be an insulator. However, the four outer-orbit electrons of the germanium atom are rather loosely bound.

Impingent thermal energy due to room temperature is sufficient to dislodge some of the shared orbital electrons. Under such circumstances, the presence of an electric field imparts a directional drift to such electrons much as in an ordinary conductive metal. We see that the dislodged electrons are effectively free, mobile charge carriers. At room temperature, however, relatively few electrons are torn loose from their covalent bonds within the crystal lattice.

Consequently, germanium cannot be classified as a conductor along with such metals as copper or aluminum, nor can germanium be considered a good insulator. Accordingly, it is termed a semiconductor. Other semiconductors are crystalline carbon (diamond), crystalline tin, and silicon.

At lower than room temperatures fewer orbital electrons are freed for conduction from their covalent binding force, and a semiconductor behaves more like an insulator. At elevated temperatures, many electrons are freed from their covalent bonds, and conduction becomes relatively good. This is only part of the story however; interestingly enough, although no mobile ions exist at any time, there is another charge carrier in semiconductors besides freed electrons.

Conduction by Holes. The nature of the covalent bond in germanium is such that an electron dislodged from its lattice leaves behind an electrical "vacuum," or "hole," to which other covalent electrons from other atoms are attracted. This being the case, an electron from an adjoining atom jumps into this hole. In so doing, this electron creates a similar hole in *its* lattice structure, which in turn is likewise filled by yet another electron from another atom. In this way there is a migration of holes toward the negative electrode and a migration of electrodes toward the positive electrode (see Fig. 2-4). Electrons which escape holes are collected at the positive electrode, whereas holes which escape electrons are attracted to the negative electrode. The important point we are making is that the holes are as instrumental in providing electrical conduction as the electrons. It is helpful to think of the holes as *positive* charges.

This prompts us to recall a somewhat analogous situation in conducting liquids (electrolytes), wherein positive ions and

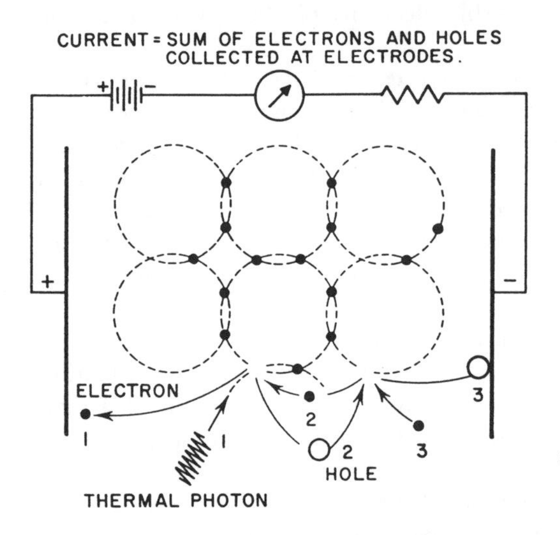

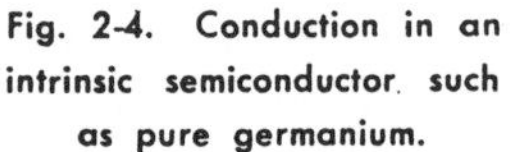
Fig. 2-4. Conduction in an intrinsic semiconductor such as pure germanium.

negative ions moved in opposite directions, *both* contributing to the net electric current. The total conduction in germanium is similarly the sum of electron current and hole current. In pure germanium, freed lattice electrons and the resultant holes are the only charge carriers available for electrical conduction. The ions produced when an atom loses an electron are constrained in their crystal lattices. They therefore do not become mobile charge carriers.

Pure germanium is frequently called *intrinsic germanium,* and the attendent electric conduction is called *intrinsic conduction.* Conduction is said to be due to generation of electron-hole pairs by thermal energy (see Fig. 2-5).

Population Turnover of Charge Carriers. Just as the formation of ions in an electrolyte is a dynamic process in which an equilibrium exists between ion formation and ion recombination, the electron-hole pairs in a semiconductor likewise undergo constant generation and recombination. When an electron or hole recombines, it is no longer available for conduction. The average time during which these charge carriers are available for conduction is called their *lifetime.* The conductivity of a semiconductor is not only proportional to the number of electron-hole pairs generated but also to their lifetimes.

The Effect of Certain Impurities in Germanium. We now arrive at a very interesting phenomenon involving the conductivity of semiconductors. Let us add a very small trace of an element with *five* outer-orbit electrons, such as antimony, to our crystalline germanium. We will form such an alloy while both substances are in their molten state, thereby permitting the antimony to diffuse uniformly throughout the germanium. After slow cooling the crystalline state of the germanium is attained. Let us see

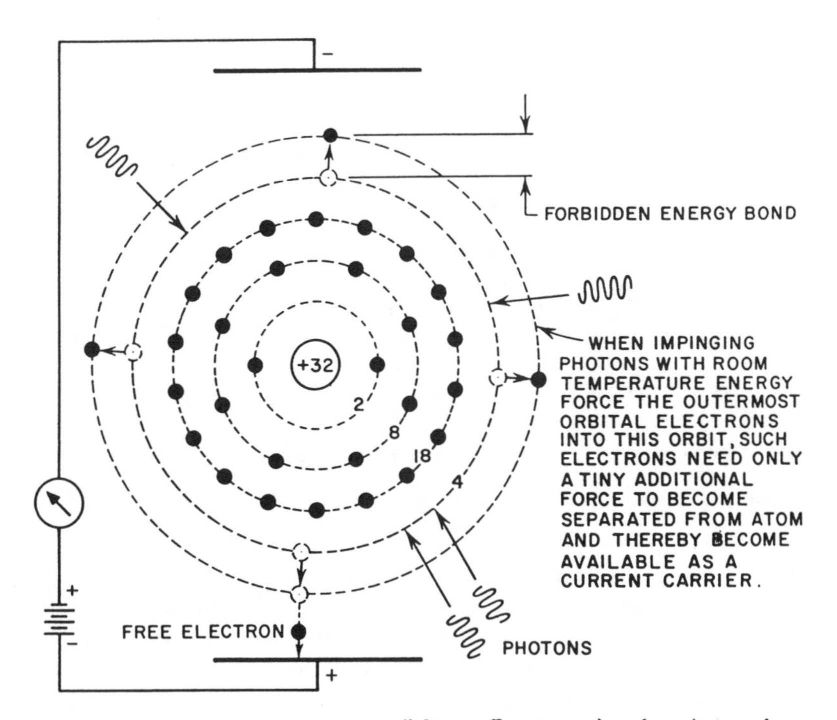

Fig. 2-5. At room temperature, a small force suffices to produce free electrons in germanium.

how the antimony fits into the atomic lattices of the resultant material.

Antimony has five electrons in its outer orbit. Only four are required to participate in the covalent lattice arrangement. The fifth electron remains loosely bound to the antimony atom *but is not bound to any other atoms.* This extra electron in the lattice structure is *more easily* dislodged from its parent atom by thermal photons than are those germanium and antimony electrons which participate in the covalent-lattice arrangement. Conduction in the "doped" germanium is greatly enhanced by the lattice-free electrons provided by the antimony.

We note that now the major current carrier is the electron. Although there is more opportunity for holes to recombine with electrons than in intrinsic germanium, all of the holes cannot be neutralized even by the excess of electrons. This is because electron-

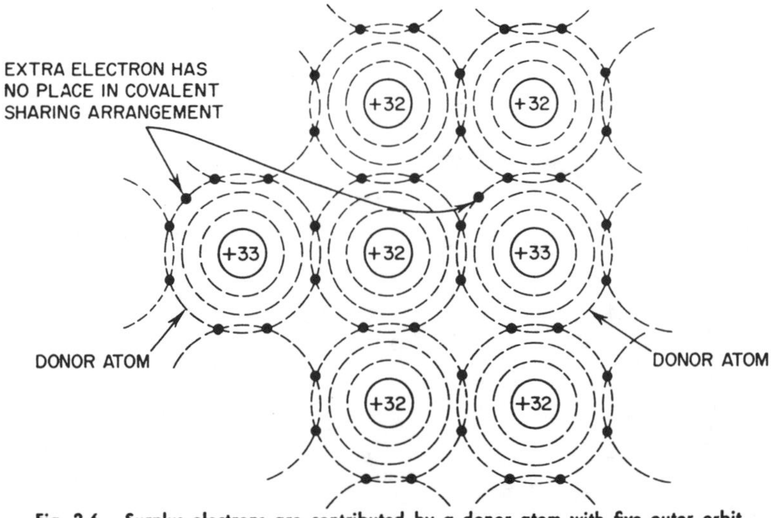

Fig. 2-6. Surplus electrons are contributed by a donor atom with five outer orbit electrons.

hole pairs are constantly being thermally generated. Even though most of the current is now the result of majority electrons, a tiny portion of the current is still contributed by the migration of minority holes. Germanium doped by an impurity which donates five electrons is known as n-type material because the majority current carrier is a negative charge. We see that impurity atoms, such as antimony, with five outer-orbit electrons donate a conduction electron in addition to four crystal-lattice electrons. Such impurity atoms are therefore known as *donor atoms* (see Fig. 2-6).

Enhancement of Hole Conduction. Another type of adulteration of intrinsic germanium is of equal importance in transistor action. If the added impurity atoms have *three* outer orbital electrons, the atomic lattice structure in some instances will be "shy" a covalent electron. Such a lattice will then harbor a "hole," which acts as a *positive* charge. The same kind of hole migration from atom to atom is caused by thermal energy as already described for intrinsic germanium. However, the holes released by the impurity require less energy for liberation from covalent bonds than do the holes generated in pure germanium. The conductivity of the germanium is greatly enhanced by the addition of such impurity atoms (see Fig. 2-7). We note that the hole is now the major current carrier.

Although greater opportunity for recombination of electrons exists than in intrinsic germanium, all of the conduction electrons are not neutralized by the excess of holes. This is because electron-hole pairs are constantly being thermally generated. Even though most of the current is now the result of holes, a tiny portion of the current is still contributed by the migration of electrons. Germanium doped by an impurity which supplies three electrons is known as p-type material because the majority current carrier is effectively a positive charge. We see that impurity atoms with three outer-orbit electrons accept electrons from the crystalline environment in order to complete the covalent bond. Such impurity atoms are therefore known as *acceptor atoms.* Indium and aluminum are examples of elements which can provide *acceptor atoms* (see Fig. 2-8).

THE P-N JUNCTION

It is now appropriate that we investigate the important characteristics of the p-n junction. As the name implies, we are dealing with phenomena occurring at the meeting zone of oppositely doped semiconductor material.

Let us visualize a bar of crystalline germanium in which half the length is of the p type, and half is of the n type material. If the positive lead from a battery is connected to the *p* electrode,

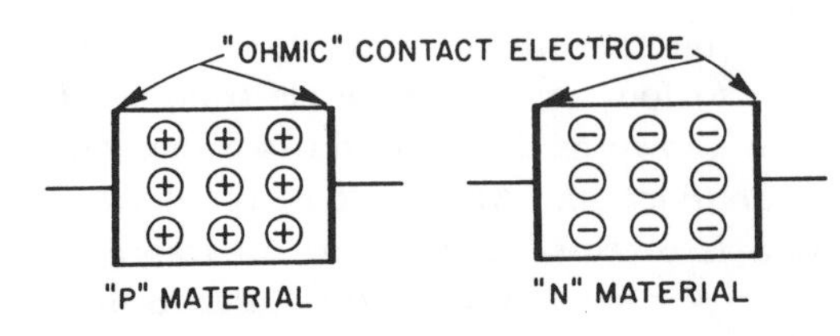

Fig. 2-7. Majority current carriers in "doped" semiconductor wafer. Conductivity in both devices is greatly enhanced over that of intrinsic conduction.

making contact with the *p* material, and the negative lead is connected through an ammeter to an electrode contacting the *n* material, heavy conduction will be found to exist. Conversely, if we reverse polarity, very little current flow can be detected.

Although the mere knowledge of this fact suffices to enable us to employ the p-n-junction diode as a rectifier, we have arrived at a point in our discussion where it is very important that we understand causes as well as effects; otherwise transistor theory cannot be evolved. After we have acquainted ourselves with the mechanism of rectification in the p-n-junction diode, it is comparatively easy to extend these concepts to embrace transistor action.

The Nature of Junctions. Inasmuch as a p-n junction is essentially a reservoir of opposite charges paired off side by side, we might expect that such a condition could only be of transitory

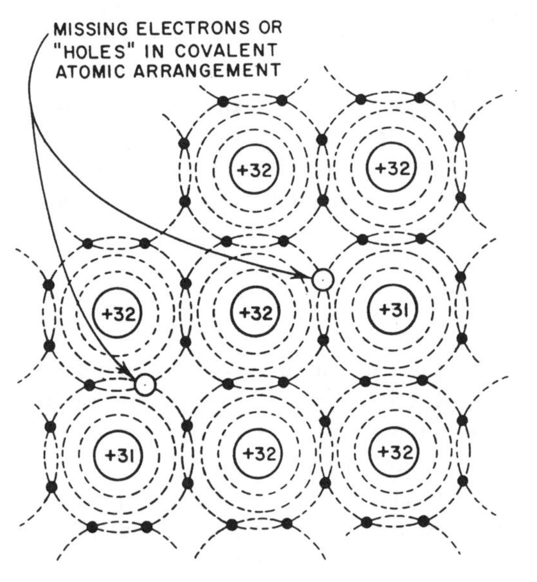

Fig. 2-8. "Holes" or loosely bound positive charges are contributed by acceptor atoms with three outer-orbit electrons.

nature, for it would appear that opposite charges must rush together, terminating their lifetime by recombination. Even if this did not happen, we might suppose that the p-n diode would behave as a charged capacitor (or an electromotive cell) which would provide a current through an external load connected to the electrodes of the diode. By investigating the reasons no such current is actually available, we can gain considerable insight into mechanisms basic to transistor action.

The Depletion Layer. Although the transition from the p to the n region may be relatively abrupt, there is always a narrow region separating the two "doped" regions. This separation zone is known as the *depletion layer.* The depletion layer has the con-

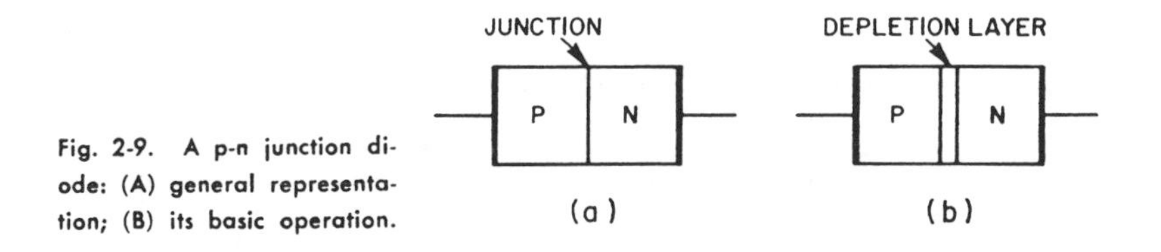

Fig. 2-9. A p-n junction diode: (A) general representation; (B) its basic operation.

duction properties of intrinsic (undoped) semiconductor material. A simple way to account for this is to postulate that in the exact middle of the p-n diode equal numbers of oppositely charged impurity atoms exist, thereby producing net cancellation of such charges (see Fig. 2-9).

A better approach involves a dynamic rather than a static concept of charge behavior. Imagine that the p-n junction has just been formed. There is indeed a tendency for the electrons and holes provided by the impurity atoms to diffuse across the junction and terminate their existence by recombination. This begins, but not without a secondary effect. The impurity atoms which provide these migratory electrons and holes are left in an ionized state, bearing the *opposite* charge from the departed carrier. However, the atoms of the impurity substances in the p and n regions are *not* mobile. Rather, like the germanium atoms, they are constrained in their positions in the crystal lattice. These impurity ions form parallel rows or "plates" of opposite charges facing each other across the depletion layer.

If a majority carrier attempts to cross the depletion layer, it will suffer one of two fates. Either it will be captured by an impurity ion of opposite sign, that is, the row of impurity ions guarding its own region, or it will get through to the depletion layer wherein it will be repelled by the opposite row of impurity ions. In the second instance it has a small lifetime which abruptly ends by recombination with a majority carrier of opposite sign

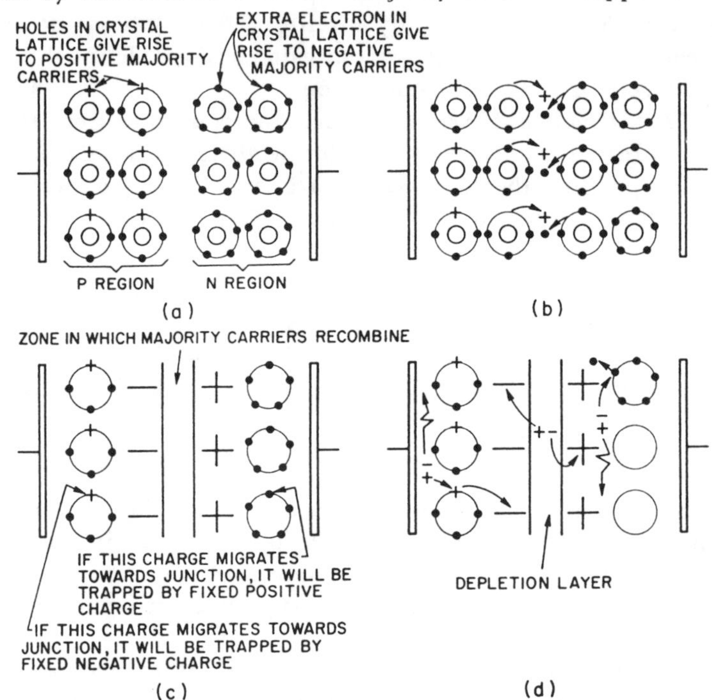

Fig. 2-10. Formation and maintenance of depletion layer is dependent upon immobility of impurity ions and continuous charge replenishment by thermally generated minority carriers: (A) two regions with acceptor and donor impurity atoms; (B) initial tendency of two charge regions is to move towards one another; (C) minus signs indicate fixed negative charges produced by loss of hole from crystal lattice, and plus signs indicate fixed positive charges produced by loss of extra electron in crystal lattice; (D) The formation of the depletion layer is a dynamic process in which charge replenishment is constantly being made by thermally generated minority carriers.

which has similarly entered the depletion layer from the oppositely doped region of the diode.

All this time, the facing rows of oppositely charged ions are themselves recombining and reforming. The whole process is one of turbulance and chaos as far as individual charge carriers are concerned, but has an aggregate semblence of order in relation to the net effect of all participating carriers. The over-all result is the *establishment* and *maintenance* of the depletion layer and the prevention of mass annihilation of mobile charges. Instead, a limited rate of majority-carrier recombination prevails in the depletion layer; more correctly, the depletion layer exists by virtue of this limited recombination. Meanwhile, in the body of both doped regions charge replenishment of ionized impurity atoms is constantly provided by thermally generated minority carriers (see Fig. 2-10).

Equilibrium of Voltage Sources. Inspection of Fig. 2-11 will now show why no external voltage or current is detectable at the diode terminals. The equivalent circuit representing the three charge regions of a p-n diode consists of three like-voltage batteries connected in series and polarized so that no voltage or current is available at the "terminals." In the equivalent circuit we note that an "internal" voltage *does* exist.

CONDUCTION IN P-N JUNCTION DIODE

Forward Conduction. Let us now connect a battery and ammeter to the p-n diode, polarization being in the so-called forward direction with the p region positive relative to the n region (see Fig. 2-12*B*). The majority carriers in both regions are repulsed from the respective electrodes and driven toward the depletion layer. Due to the energy acquired from the electric field, such carriers cross over into oppositely doped regions and are collected at the electrodes there. Some recombination occurs in the depletion layer; but many of the charge carriers, because they now have energetic and directed motions, pass one another, continuing their journeys until arrival in opposite conductivity regions. Such arrival manifests itself as current flow in the external circuit, much as plate current in a vacuum tube results from electrons arriving at the positively charged plate.

Relative Doping. It is not at all necessary, nor even desirable that the doping in the two impurity regions of a p-n diode be uniform. It is possible to have conduction achieved preponderantly by one carrier, either electrons or holes. For example, a p-n-diode

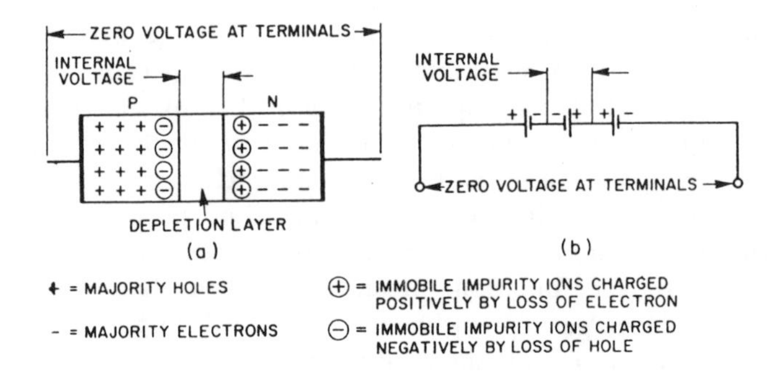

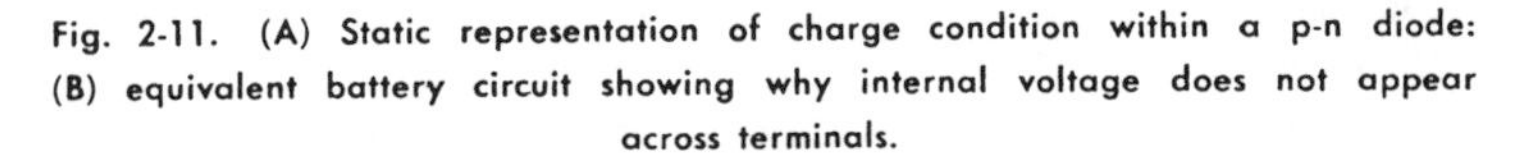
Fig. 2-11. (A) Static representation of charge condition within a p-n diode: (B) equivalent battery circuit showing why internal voltage does not appear across terminals.

rectifier may be made with relatively heavy impurity concentration in the p region as compared to the n region. In such a case, forward conduction is primarily achieved by holes because of their availability. Different semiconductor devices and different applications have their unique requirements for relative concentration in the two impurity regions.

Reverse Conduction. Next, we reverse the polarity of the battery connections, thereby polarizing the p region negative with respect to the n region (see Fig. 2-12C). Mobile majority charges in the p and n regions now are attracted to opposite charges appearing at their respective electrodes. Thus, electrons in the n region recede from the junction and move closer to the electrode contacting the n region. The mutual action of the charges in the

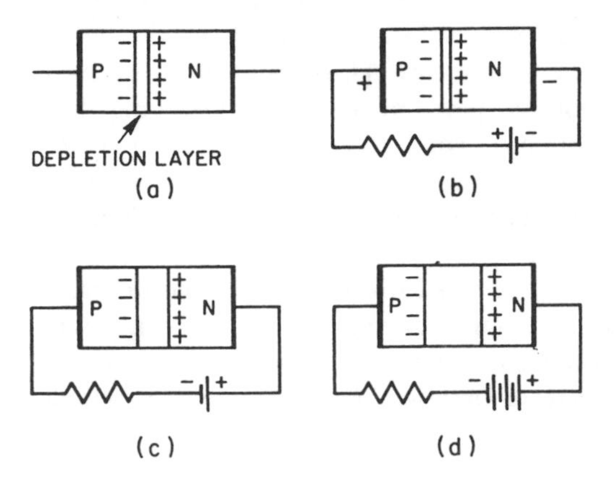

Fig. 2-12. Variation in width of depletion layer with applied voltage: (A) no voltage applied to diode; (B) forward conduction voltage; (C) low-reverse voltage; (D) higher reverse voltage.

oppositely doped regions effectively widens the depletion layer, thereby decreasing the effective capacitance of the junctions. If we raise the voltage, the widening of the depletion layer is increased (see Fig. 1-12*D*). In this way, a junction diode behaves as a voltage-controlled variable capacitor in which the "plates" of the capacitor are made up of the oppositely charged impurity ions facing the depletion layer. The "dielectric" of this capacitor behaves as intrinsic germanium because of the near-absence of majority carriers.

We note that current cannot flow by transport of majority carriers because the electric field tends to pull them away from the depletion layer. A tiny current *does* flow due to thermally generated electron-hole pairs within the depletion layer. This is essentially the current which would flow in pure germanium. Thus the dielectric is a somewhat leaky one because germanium is a semiconductor not an insulator.

EFFECT OF REVERSE VOLTAGE ON CURRENT. The small reverse current in junction diodes is nearly independent of applied voltage over a wide voltage range (see Fig. 2-13). Let us see why this is so.

We know that in a vacuum-tube diode, the electrons are accelerated more as the plate-cathode voltage is raised. This results in a greater number of electrons to reach the plate in a given time, which is tantamount to higher plate current. In contrast to this situation, the charge carriers in the depletion layer of a p-n junction polarized for reverse conduction reach their respective electrodes at constant rates. It is true that the electric field tends to accelerate the charges as in a vacuum diode. However, when the electrons and holes are forced to move faster, the increased number and violence of collisions and reflections from lattice atoms offset the initial increase in speed of carrier movement. The net rate of arrival of carriers at their respective electrodes therefore remains approximately constant despite the initial accelerating force of the electric field.

EFFECT OF TEMPERATURE ON REVERSE CURRENT. Although reverse current in the p-n diode is not greatly affected by wide variation in the applied voltage, this current increases extremely rapidly with temperature. At first consideration, this may not appear consistent with the situation described with respect to voltage variation. It might be argued that, to a first approximation, the over-all effect should be about the same whether the carriers are under impulsion of an electric field or are being provoked by thermal photons. In either case, when their motion becomes more violent, net increased speed of movement towards the electrodes is counteracted by increased, and more violent, encounters with atomic forces.

It might further be anticipated that the increased vibration of atoms within their lattices would impart even more violent deflection to passing carriers and further impede their general drift toward their respective electrodes. This would constitute a factor operating in behalf of a current *decrease* with respect to

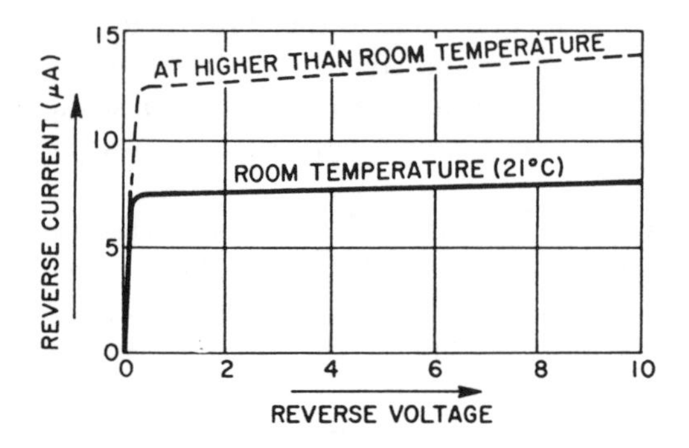

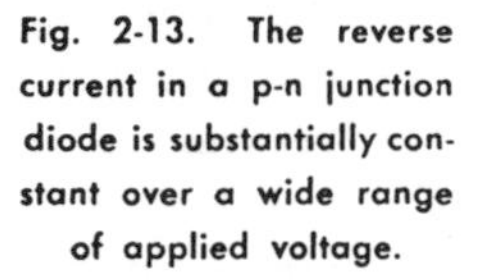
Fig. 2-13. The reverse current in a p-n junction diode is substantially constant over a wide range of applied voltage.

increased temperature. Although such arguments are valid in themselves, they are completely overwhelmed by another consideration: to wit, the tremendous increase in the *number* of electron-hole pairs resulting from small increases in temperature.

Thus, the sheer *availability* of charge carriers is much more significant in the over-all picture than the additional road blocks created to impede their drift toward their respective electrodes.

Reversion to Intrinsic Conduction at High Temperatures. If the temperature of a semiconductor junction diode exceeds room temperature by too many degrees, the reverse current can become as great as, or greater than the forward current, for given voltage values. At high temperatures conduction in both the reverse and forward direction is primarily due to thermally generated carriers in the parent material. In such a case, the impurity atoms are not of any great consequence, and conduction is nearly the same for both polarities of applied voltage. Rectifying action is thereby lost as the material assumes the characteristics of its intrinsic state.

Effect of Temperature on Rectification. At depressed temperatures, the reverse current becomes very low because of the diminished availability of thermally generated carriers. This can have both desirable and undesirable effects in semiconductor devices. In diode rectifiers low reverse conduction is generally advantageous. However, it is also true that forward conduction suffers at low temperatures. Rectification efficiency is governed by the ratio of forward-to-reverse conduction. This implies that rectification efficiency drops off at both low and elevated temperatures.

In a transistor, operation at very low temperatures may cease in a manner roughly analogous to a vacuum tube deprived of thermionically produced electrons, as might happen by reduction of filament current. We see that p-n-junction rectifiers must operate within an appropriate temperature range, and we can anticipate this to likewise be the case with transistors which embody p-n diodes in their structure.

Excessive Reverse Voltage. Although the current in a reverse-biased p-n junction remains at a low saturation value over a wide range of applied voltage, a sufficiently high reverse voltage projects the operation of the diode into an entirely different region. When a certain value of voltage is reached, the current abruptly undergoes a tremendous increase.

We are reminded here of the phenomenon of ionization in a gaseous diode. The sudden breakdown is nondestructive as long as a series-limiting resistor is incorporated to restrict the maximum allowable current. The voltage at which breakdown occurs is governed by temperature, by the impurity concentrations in the doped

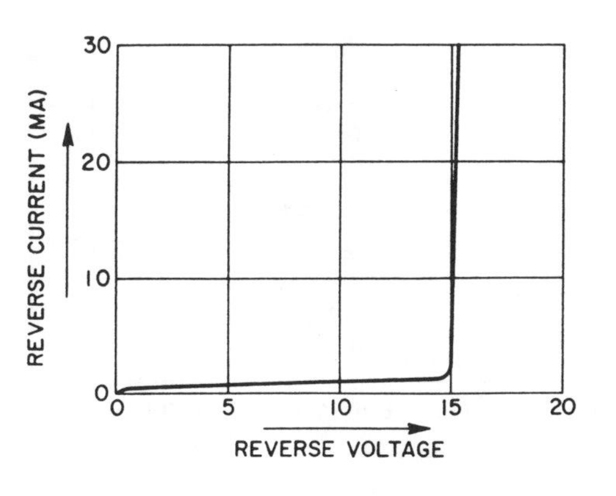

Fig. 2-14. Voltage breakdown characteristic in a p-n junction diode.

regions, and by geometrical factors in the structure of the diode. Breakdown potential from several to 100 volts is readily attainable.

When operated at breakdown voltage, the reverse-biased p-n diode performs the useful function of voltage regulation in a circuit similar to that used with gaseous regulators (see Fig. 2-14). An advantage, in addition to selection of a wide breakdown range, is that the firing and operating voltage of the reverse-biased diode is substantially the same.

Voltage-Breakdown Phenomena. Two different breakdown mechanisms are postulated. One theory accounts for the sudden increase in current in terms of an avalanche phenomenon similar to the Townsend effect in electrified gas.

AVALANCHE BREAKDOWN. According to this theory the thermally generated charge carriers in the depletion layer obtain sufficient energy from the accelerating force of the electric field to dislodge other carriers from their covalent bonds. A cumulative production of carriers is thereby produced, for each available carrier impacts one or more lattice-bound charge with sufficient force to overcome the covalent bond. Such impacted charges not only become current carriers but participate in the avalanche process by impacting other lattice-bound charges. A tremendous number of carriers are created, and conduction becomes exceedingly high (see Fig. 2-15).

ZENER BREAKDOWN. A second theory, that formulated by Zener, accounts for the current increase by assuming that the force

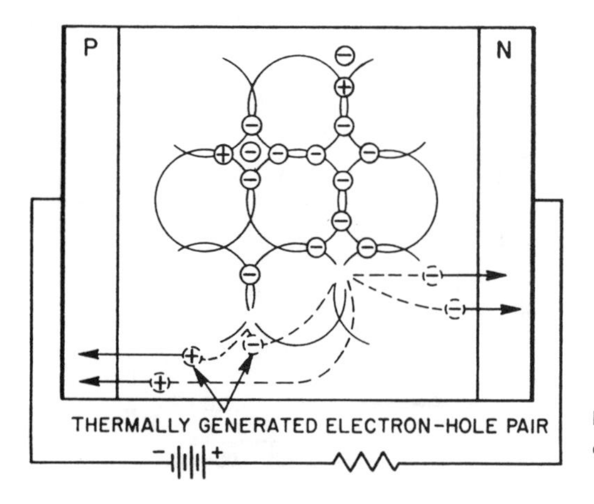

Fig. 2-15. Avalanche breakdown in p-n junction. Impact process is cumulative.

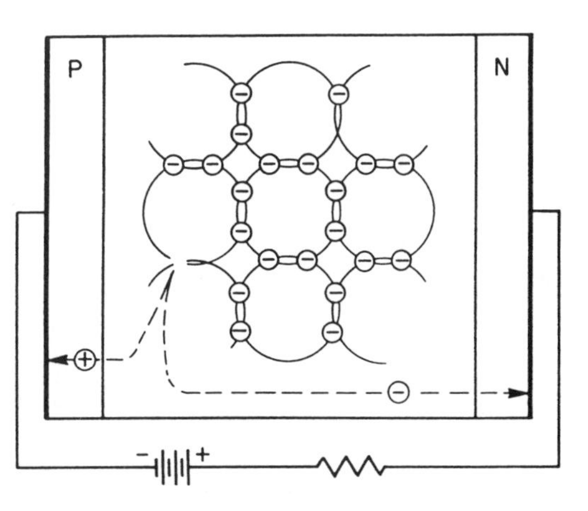

Fig. 2-16. Zener breakdown in p-n junction. Electrons are torn out of their covalent bonds by the force of the electric field.

of the electric field actually tears loose a large number of charges from their covalent bonds in the atomic structure of the crystal lattice (see Fig. 2-16). The difference between the two theories is a somewhat subtle one insofar as practical results are concerned. The Zener effect predominates at low voltages, below approximately 5½ volts in silicon. In germanium, it appears that avalanche breakdown occurs first. Therefore, except under special conditions, Zener breakdown is not encountered in germanium.

So-called *Zener diodes* are silicon junction diodes which exhibit Zener breakdown below 5½ volts and avalanche breakdown for higher voltage. There is a natural temptation to harness the carrier multiplication due to either of these breakdown mechanisms as a current-amplifying process in transistors. In a subsequent chapter we will encounter a transistor specially designed to utilize breakdown phenomena.

3. transistor action

We have discussed basic principles of conduction in semiconductors. By means of a unique application of these principles it is possible to create the semiconductor amplifier known as a *junction transistor.* The junction transistor is both physically and conceptually an extention of the p-n-junction diode. Whereas the p-n-junction diode comprises *two* oppositely doped regions separated by a junction or depletion layer, the junction transistor contains *three* alternately doped regions separated by *two* depletion layers. In place of the p-n diode, we have the p-n-p or n-p-n structures. It will be shown that the operating principles of the two structures are essentially similar. The n-p-n structure derives its characteristics primarily from the effects of electron conduction. The converse is true with respect to the p-n-p structure, which depends primarily upon the effects of hole conduction.

From the standpoint of external circuits, the n-p-n transistor is somewhat more analogous to vacuum-tube circuits due to similar polarization of B battery connections. Consequently the n-p-n transistor will receive favored treatment in subsequent analysis except where it is directly expedient to discuss its counterpart, the p-n-p transistor.

VACUUM-TUBE ANALOGY

The Positive-Grid Vacuum Tube. A vacuum tube with its grid operated slightly positive with respect to its cathode provides us with an analogous structure for discussion of transistor action.

(The basic amplifying action of vacuum tubes is generally not degraded by such operation; furthermore some tubes, such as those making use of a "space-charge" grid, actually do operate with positive grids.) The cathode and the positive grid of a tube constitute a rudimentary electron gun. Such an electron gun projects electrons in the direction of the collector electrode, or plate. By a slight modification of the p-n diode structure, we can create a semiconductor "gun" which also projects charge carriers into the vicinity of a collector electrode. Such a semiconductor gun is depicted in two forms in Figs. 3-1 and 3-2. The "gun" of Fig.

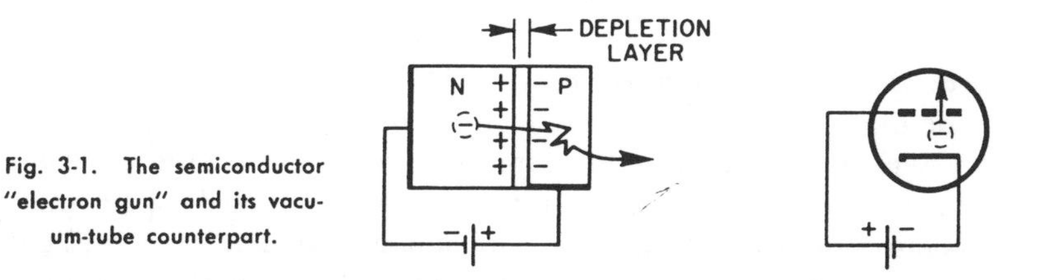

Fig. 3-1. The semiconductor "electron gun" and its vacuum-tube counterpart.

3-1 projects electrons into an n-p-n transistor, whereas the gun in Fig. 3-2 projects holes into a p-n-p transistor. The semiconductor guns differ from the p-n diode in two essential respects. First, one electrode contacts the *side* rather than the end of the structure; secondly, one doped region is very narrow. Let us consider the result of biasing the n-p gun in the forward direction.

Charge Injection. As with other junction diodes, forward conduction narrows the depletion layer and injects majority carriers from one doped region to the other, wherein the opposite charges assume minority current-carrying status. Assume in this case that the doping of the n region is much heavier than the doping in the p region. This causes the predominant portion of charge movement to consist of electron injection from the n to the p region. We should keep in mind that the forward-bias voltage applied to the emitter-bias diode effectively lowers the "built-in" internal voltage which exists across the depletion layer. This permits the flow of electrons, or, as we say, their "injection," from the emitter region to the base region.

It is of paramount importance to appreciate that the electric field gradient wthin the base region itself is very small. Once the electrons have been injected into the base, they no longer are subjected to an appreciable electric field. Instead, their subsequent movement is derived from the effects of thermal agitation. Due to the fact that like charges repel, there is a tendency for the electrons to continually seek regions of the base material in which the electron population is less than in their present region. This

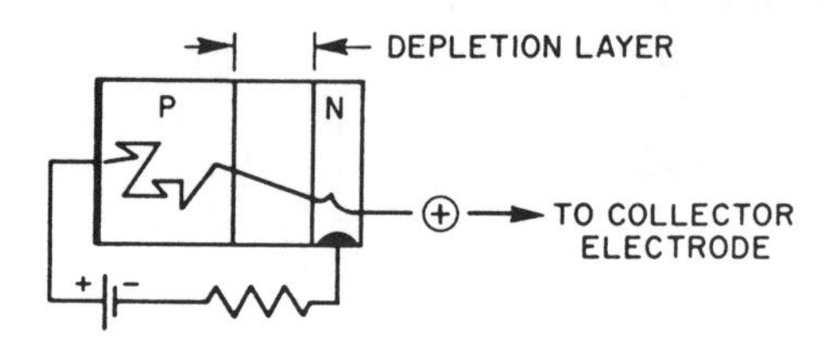

Fig. 3-2. The "hole" gun.

gives rise to the so-called *concentration gradient* and serves to guide the net movement of electrons across the thin base region.

We see that, due to the concentration gradient, the electrons have a directed transverse movement superimposed upon their random motions. Because the p region is very *thin,* the injected electrons arrive at the far-end region of the p material rather than at the side-contacting electrode. Such electrons are then susceptible to the attractive force of a collector electrode.

The analogous situation in the vacuum-tube gun involves the electrons emerging through the apertures of the positive grid. These electrons, too, are susceptible to the polarizing field of a collector electrode (plate). Inasmuch as parallel mechanisms serve to enhance the value of analogy, it behooves us to note that the process of carrier diffusion is involved in both semiconductor conduction and thermionic emission (see Fig. 3-3). In the latter case, electrons are impelled by thermal agitation to diffuse from the interior to the emitting surface of the cathode. In our semiconductor gun, while the majority carrier is in the 'emitter" region, it is similarly impacted in random directions by the thermally induced atomic vibrations. In a very real sense, we produce a kind of a "thermionic" emission here too.

Action Within the Transistor Structure. The semiconductor electron and hole "guns" are not capable of operation actually detached from the remainder of the transistor. However, they do operate *within* the complete transistor structure as described. A second diode section, formed by the base and collector, provides the medium and means for collecting the charges injected by the

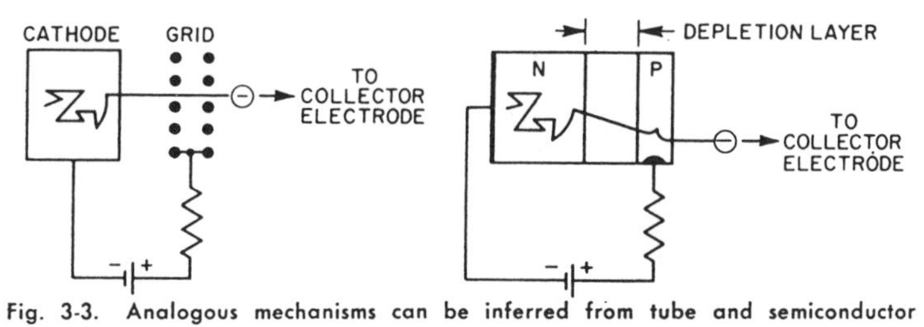

Fig. 3-3. Analogous mechanisms can be inferred from tube and semiconductor "electron gun."

"gun." The first diode, formed by emitter and base regions, is *forward*-biased. The second diode section, formed by base and collector regions is *reverse-biased.*

We see that the structure suggests two back-to-back diodes, one forward-biased, the other reverse-biased (see Fig. 3-4). A significant feature possessed by the transistor is, however, lacking in such an arrangement of physically separate diodes. We note that once majority carriers are emitted from the emitter region, they become minority carriers in the base and collector regions. This is true in the base region by virtue of reversal of the doping and remains true in the collector region for a different reason, i.e., the application of reverse bias (see Figs. 3-5 and 3-6).

Power Amplification in the Transistor. In the transistor, as in the circuit comprising physically separate diodes, the reverse-biased diode conducts very feebly. The small conduction which exists is due to thermally generated carriers in the depletion layer. In the separate-diode arrangement, this condition is *not* affected by conduction phenomena in the independent forward-biased diode. In the transistor, however, a different situation prevails, for here the forward-biased diode constitutes our previously discussed charge gun. When the forward-biased diode is carrying current, the majority carriers responsible for this current are injected into the depletion layer of the reverse-biased diode, becoming therein

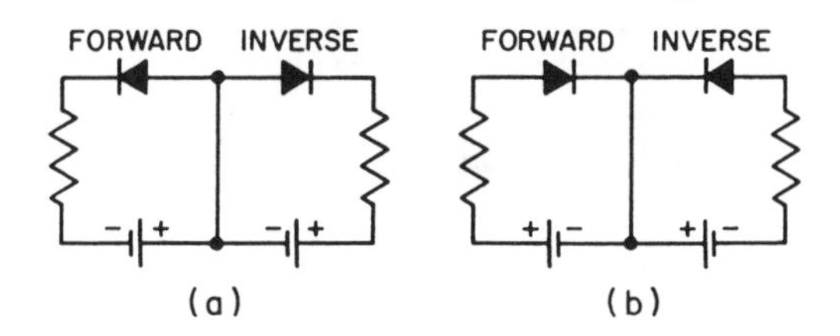

Fig. 3-4. Equivalent diode circuits which depict biasing polarities of transistors: (A) circuit illustrating biases in n-p-n transistor; (B) circuit illustrating bias in p-n-p transistor.

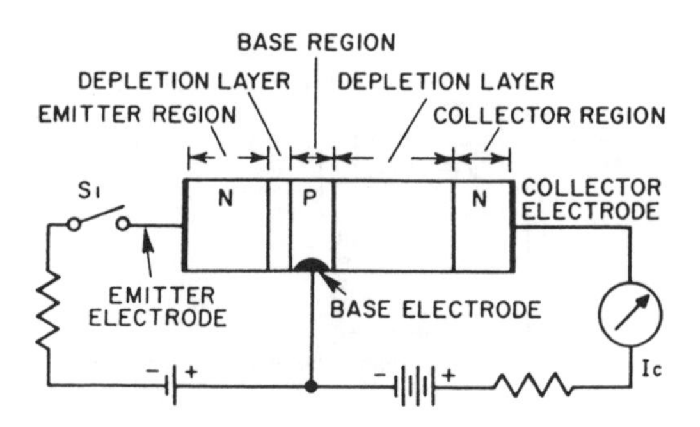

Fig. 3-5. Complete transistor structure with a possible biasing arrangement. Basic transistor action is demonstrated by this circuit. Collector current flows when switch S_1 is closed, but falls to very low value when S_1 is open.

a rich source of minority carriers. As a result, very nearly the same current flows in the collector circuit as in the emitter circuit. Here at once we have the derivation of the name *transistor* and the secret of the device's ability to provide power amplification. We see that a current flow in the relatively low-resistance emitter-base diode causes, by charge injection, almost the same current to flow in the relatively high-resistance base-collector diode. A given current evolved in a high-resistance source furnishes more power than the same current associated with a low resistance. This is simply a manifestation of the relationship derived from the Ohm's law, namely $P = I^2R$.

Emitter-Base Doping. In order to achieve good transistor action, the emitter-base diode must be doped in such a way as to enhance injection of the charge which behaves as a minority carrier in the reverse-biased base-collector diode. Thus, in the n-p-n transistor we are interested in injecting electrons into the base-collector diode. However, the forward-conductive component of current in the emitter-base diode which is due to passage of holes from base to emitter does not serve this purpose. This part of the

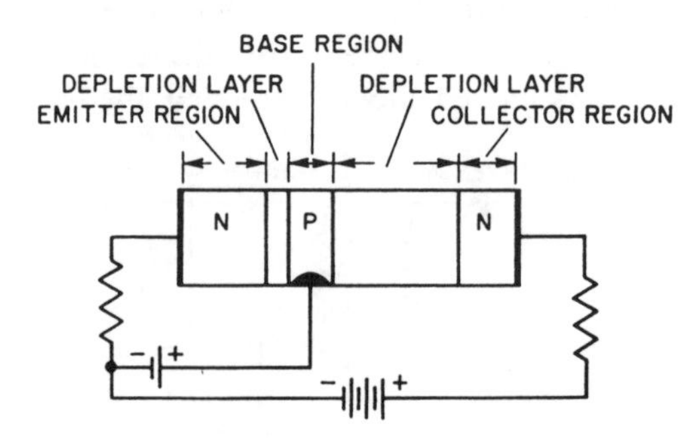

Fig. 3-6. An alternative method for biasing the transistor.

forward current degrades the power-amplifying ability of the transistor because an accompanying increase in reverse current is not thereby produced in the base-collector diode. Consequently, it is desirable to dope the emitter region, the source of injected electrons in the n-p-n structure, much more heavily than the base region. This ensures forward current conduction primarily by electron flow. (The opposite inequality in relative doping of emitter-base regions prevails in the p-n-p transistor, for in this case we wish to inject *holes* into the reverse-biased base-collector diode.)

Collector Action in the Transistor and the Vacuum Tube. Vacuum-tube action depends upon controlled injection of current carriers into a space-charge region produced by a polarized collector electrode. The space-charge region ahead of the plate

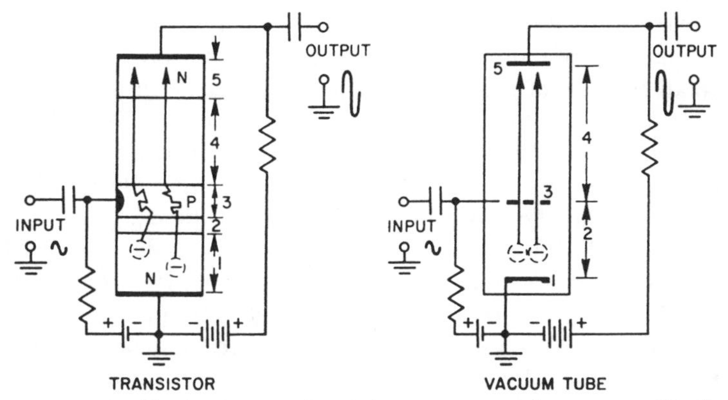

Fig. 3-7. Amplification in a transistor is in many ways analogous to amplification in a vacuum tube. Although generally not required and often undesirable, a small positive bias is shown applied to the grid of the vacuum tube in order to reinforce the analogy.

receives injected carriers (electrons) from the grid apertures. The grid can control the electron injection with negligible power consumption compared to the power represented by plate current. This accounts for the tube's ability to provide power amplification.

An analogous situation exists in a transistor (see Fig. 3-7). The complete transistor structure consists of a sandwich of alternately doped semiconductor regions. The doped regions are separated by junctions or depletion layers. Assuming we are dealing with an n-p-n transistor, we identify the first n region as the emitter, the p region as the base, and the second n region as the collector. The second depletion layer is analogous to the space between grid and plate of a vacuum tube. The collector region collects injected changes as does the plate of a tube. The base region is capable of controlling the injected changes in a way suggestive of the grid of the vacuum tube (see also Table 3-1).

TABLE 3-1: AMPLIFICATION ANALOGY OF FIG. 3-7

Transistor region	*Function or property*	*Vacuum-tube region*
1. Emitter region	Charges originate	1. Emitter or cathode
2. Depletion layer	Electric field region	2. Emitter-grid region
3. Base region	Control region	3. Grid
4. Depletion layer	Electric field region	4. Plate-grid region
5. Collector region	Charges collected here	5. Collector or plate

The signal power required by the base to accomplish such control is a small fraction of the controlled power available in the external collector circuit. This basic requisite of power amplification is brought about by the unequal doping of emitter-base regions and by the high resistivity of the reverse-biased base-collector diode compared to the resistivity of the emitter-base diode.

Collector Voltage vs. Collector Current. Considerable insight into transistor action may be attained through inspection of the relationship between collector voltage and collector current. In the graph of Fig. 3-8 we observe several interesting characteristics. First, it is quite obvious that collector current is substantially independent of collector voltage relationship much as in a pentode vacuum tube (see Fig. 3-9). In both devices, the output circuit behaves as a constant-current generator. In the transistor, the collector voltage exerts virtually no attractive force on carriers originating in the emitter region. These carriers diffuse across the thin base region and experience attractive force only after entering the depletion layer of the collector region. No matter what voltage is applied to the collector, the rate at which minority carrier collection takes place is primarily governed by current flow in the emitter-base region, the charge gun of the transistor.

Similarly, in a pentode tube, the electric field established by plate potential does not exert attractive force on electrons behind the "muzzle" of the electron gun, that is, the positive polarized screen-grid. The electrons emerging from the apertures of the screen-grid comprise the only current available to the plate. The rate of flow of these electrons is governed by the voltage existing between screen-grid and cathode.

Residual Collector Current with no Carrier Injection. We note that with zero emitter current, there is a residual collector current. Under this condition, the base-collector diode displays the characteristics of a simple reverse-biased junction diode. The small current is due to minority carriers which are thermally generated in the depletion region between base and collector. Here, likewise, the current is not governed by applied collector-to-base voltage, but rather by the mechanism which makes minority carriers available. In this case the carriers are thermally generated. Therefore the residual collector current I_{CO} is at the mercy of ambient temperature. This current known as the *collector saturation current,* or collector cutoff current, cannot be eliminated by reversing the polarity of the bias applied to the input diode, that is, to the emitter-base diode.

Negligible "Wasted" Collector Voltage. One of the impressive aspects of the collector-current vs. collector-voltage relationship is the existence of full collector current at extremely low collector voltage. This is a desirable characteristic, for it provides realization of almost the full 50% efficiency theoretically obtainable from class-A amplifiers. The physical mechanism responsible for this unique performance again involves the fact that carrier injection

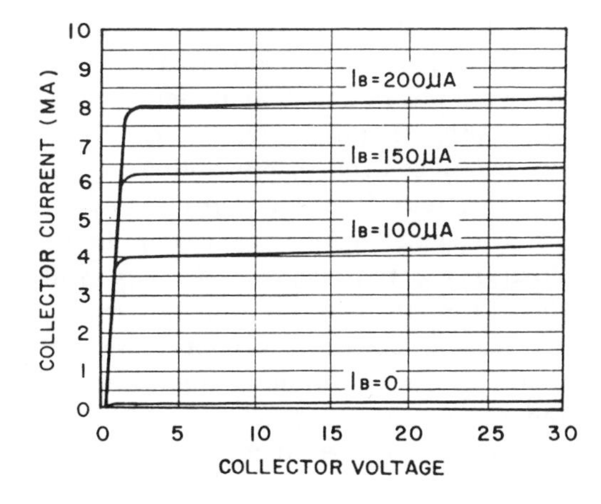

Fig. 3-8. Collector current vs. collector voltage for a typical n-p-n junction transistor (common emitter circuit). We are reminded of the pentode plate current vs. plate voltage relationship.

into the collector region is governed by the number of carriers diffusing through the thin base region, not by any attractive force established by collector voltage. The voltage applied to the collector may be considered to perform an external-circuit function; i.e., it permits selection of an operating point and load line so that the amplifier signal can undergo its peak-to-peak voltage

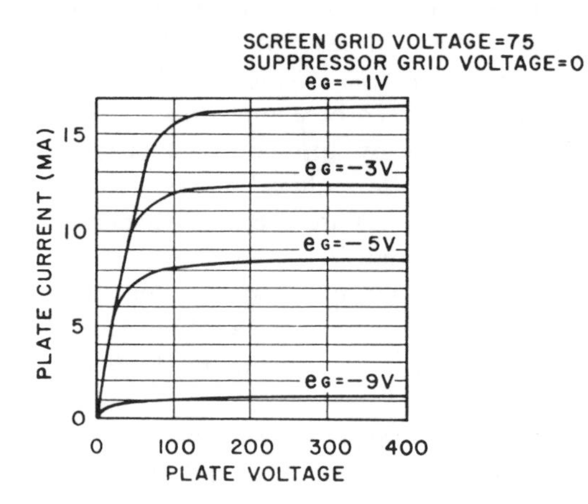

Fig. 3-9. Plate current vs. plate voltage for a typical small pentode tube. Note similarity with Fig. 3-8.

excursions. We also note that the curves are about equally spaced with respect to equal increments of emitter current. From this we deduce that amplification by transistor action is a current-actuated phenomenon wherein output current is proportionate to input current. This is in contrast to ordinary vacuum-tube operation in which a varying voltage applied to the control element causes variations in output current. However, even current-controlled amplification has its counterpart in some types of tube operation. For example, in class-B amplifiers the grid is driven into its current-consuming region. Under this condition the tube may be said to be current-controlled, resembling in this respect transistor action.

Electrode Currents in the Biased Transistor. When the transistor is polarized by its biasing batteries, the apparently paradoxical situation depicted in Fig. 3-10 exists. The emitter-base diode supports entirely different currents in its two leads. Such a situation obviously cannot exist with a simple p-n junction diode. However, in the complete transistor, this phenomenon is the consequence of the circuit connections. The large current measured in the emitter lead is the result of current circulating in *two* circuits, *not* merely the circuit directly responsible for forcing forward current through the emitter-base diode. The vacuum tube operated with positive grid gives us a good analogy of the situation. The current in the "emitter" lead of both the transistor and the vacuum tube is the sum of input and output currents. We see that opening the switch S_w in both transistor and tube circuits leaves only the input "diode" baised in both instances. Under such conditions we would obtain equal currents in the input diode leads. That is, in the transistor, 100 μa would be indicated by base and emitter ammeters; in the tube arrangement, 100 μa would be indicated by grid and cathode ammeters.

Alpha, the Current Gain Factor. We have found that the basic cause of power amplification in the transistor is the transference of current carriers from a low-resistance to a high-resistance region. The transferred current originates in the emitter region as majority

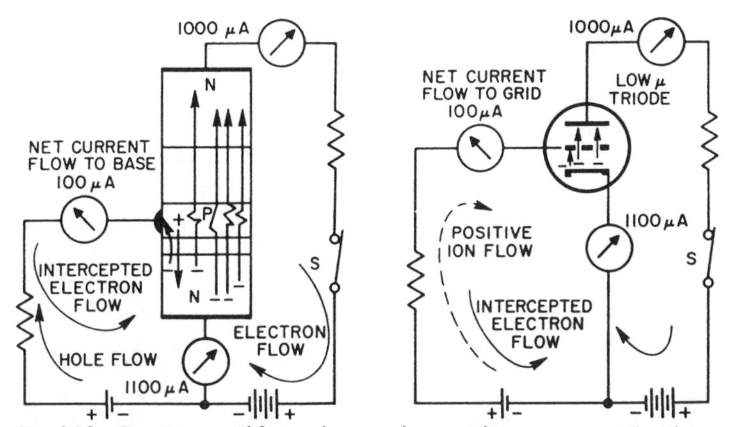

Fig. 3-10. Transistor amplifier and an analogous tube arrangement for demonstrating current in control circuit. In both circuits the current in the connection to the control element circulates in the input circuit and is lost to the collecting element. The situation for a p-n-p transistor would be the same as that shown for an n-p-n structure except for reversal of battery polarities and interchange in roles of electrons and holes.

carriers. As a result of forward-conduction bias applied to the emitter-base diode, these carriers cross the depletion layer and enter the narrow-base region. Here, due to impurity doping of opposite sign to the emitter region, the injected carriers assume the status of minority carriers. Their movement in the base region is largely random, the impetus being derived primarily from thermal energy. Unfortunately, not all of the carriers injected from emitter to base ultimately enter the electric field in the depletion layer between the two regions of the reverse-biased base-collector diode. Some of these charge-carriers suffer recombination in the base region (see Fig. 3-11). The random motions of some result in return to the emitter region.

Yet another loss in injection efficiency is due to emission of opposite-sign charges from base to emitter. For example, in the n-p-n transistor it is desired that electrons, as carriers of the for-

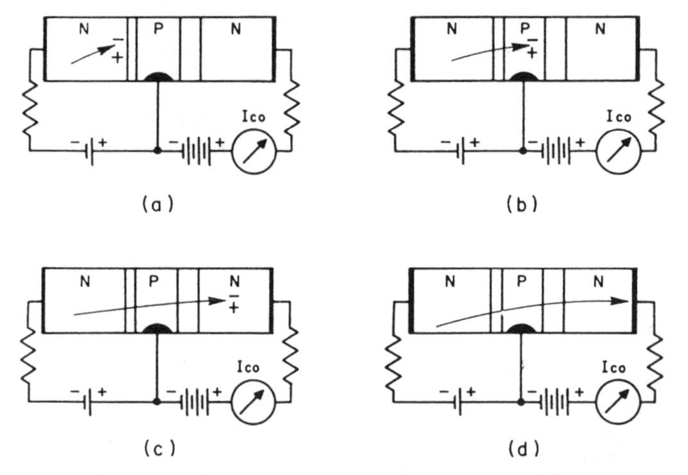

Fig. 3-11. The effect of recombination at various regions within the transistor: (A) and (B) Recombination in emitter and base regions effectively result in loss of carrier insofar as concerns collector current. Situation depicted in (B) produces undesired base current. (C) and (D) Recombination of injected carrier in collector region or at collector electrode establishes collector current. *Note:* Recombination in depletion layers would also prevent completion of injection process. Only I_{CO} would flow in the collector circuit.

ward-conduction current, be driven from the emitter to the base region. However, there is a second component of forward conduction which yields no profit insofar as contributing towards a "space" current effectively flowing from emitter to collector. It is the movement of holes from base to emitter which constitutes this loss. We recall that in an n-p junction, a small portion of forward conduction current is due to the flow of minority carriers in the opposite direction to the main movement of majority carriers. The two effects are additive insofar as concerns current flow in the external circuit. In the transistor, we designate a figure of merit α *(alpha)*, which is simply the ratio of input to output current measured in the common-base circuit.

$$\alpha = \frac{\text{collector current}}{\text{emitter current}} = \frac{I_2}{I_1}$$

If none of the aforementioned injection losses existed, the value of α would be 1. Because not all injected carriers reach the collector and because of reverse injection of minority carriers, α's between 0.90 and 0.99 are typically found in actual junction transistors. We note that the transference of current from a low- to a high-resistance region makes possible power gain notwithstanding current gains less than unity. Furthermore, we will later find that current gains greatly in excess of unity are attained through use of circuit configurations other than the grounded base.

α is unique to transistors; no counterpart is found in commonly used vacuum-tube gain parameters such as μ or gm. We can, however gain further insight into the nature of α by devising a special tube arrangement to which an analogous α may be assigned to describe a similar figure of merit as that prevailing in transistors. In Fig. 3-12 the emitted carriers (electrons) do not all reach the plate. Some are intercepted at the grid wires. Such intercepted electrons produce grid current and are analogous to the electrons which suffer recombination in the base region of the n-p-n transistor. If the tube is slightly gassy, a positive ion flow from grid to cathode adds to the grid current but does not add to plate current. This flow of positive ions from grid to cathode suggests the base-to-emitter flow of holes in the n-p-n transistor. Thus, cathode current exceeds plate current, and α expresses the ratio of "collected" to "injected" current.

$$\alpha = \frac{\text{plate current}}{\text{cathode current}} = \frac{I_2}{I_1}$$

Improving Carrier Injection in Tube and Transistor. In the tube, current interception by the positive grid can be reduced by employing a grid structure made of fine wire and wound with .atively large spaces between adjacent wires. In the ransistor, similar objective is achieved by doping the base region less .eavily than the emitter region. The relatively sparsely doped base then offers fewer recombination opportunities to the injected charges. It furthermore injects fewer minority charges into the emitter region.

DIFFERENCES BETWEEN TUBE AND TRANSISTOR

Nonanalogous Mechanisms. In establishing an analogy between tube and transistor operation, we must be fair with ourselves and bring to light basic differences as well as parallel mechanisms (see Fig. 3-13). An important case in point involves the driving force which causes charge carriers to traverse the charge gun of the two devices. For the transistor, we concern ourselves with the way in which emitted charges pass to and through the base region. The electric field due to the inverse bias

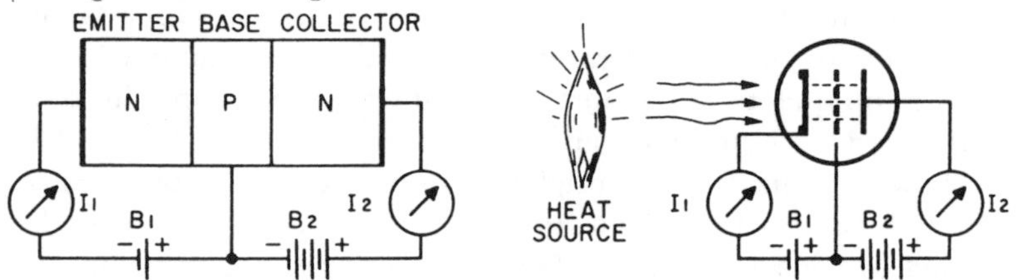

Fig. 3-12. A vacuum-tube analogy for α in the transistor. Here we produce cathode emission with the aid of an external heat source. Space current is then governed by positive grid voltage supplied by battery B_1.

applied to the base-collector diode exists as a potential gradient across the depletion layer separating base and collector regions. This field does not appreciably extend into the base region or into the emitter region. Consequently, charges emitted into the base region do not experience an attractive force from the collector region. Some of these charges are lost by recombination with charges of opposite sign in the base region, this constituting collection by the base electrode. Still others return to the emitter-region. Because the base region is very thin, most of these charges are impelled by thermal agitation to zig-zag into the base-collector depletion layer. Here they are swept up by the strong electric field prevailing in this region and collected in the collector region. We see that the charge must have suffered relatively long delay in crossing the base region. Indeed, it is here that we encounter one of the basic limiting factors to high-frenquency operation.

In contrast to the relatively slow charge transit imposed by thermal diffusion in the base region of transistors, the emitted charges in the vacuum tube are constantly being accelerated by a practically continuous electric field. Furthermore, the emitted charges leave the cathode and are not thereafter buffeted or impeded to any extent by constituents of matter or their force fields. Thus, the factor of charge transit time is a more favorable one in designing tubes for high frequencies than is the case with transistors.

Other Differences Between Transistors and Tubes. Another difference between basic mechanisms in transistors and tubes is the absence of a space charge surrounding the emitter region in the transistor. In the tube, there is generally an accumulated cloud of electrons or "space charge" which forms in front of the cathode and limits emission. Inasmuch as our discussion generally deals with tubes having positively biased grids, the space charge can be considered neutralized. This was, indeed a factor in postulating positive grid operation as a close analogy to transistor action.

A final distinction between transistor and tube operation, one which bears significantly on biasing requirements is the fact that a component of the collector current, the so-called *collector saturation current* or collector cut-off current, is not subject to control by the emitter-base diode. Whereas the plate current of a vacuum tube can be reduced to zero by sufficient cutoff grid bias, the

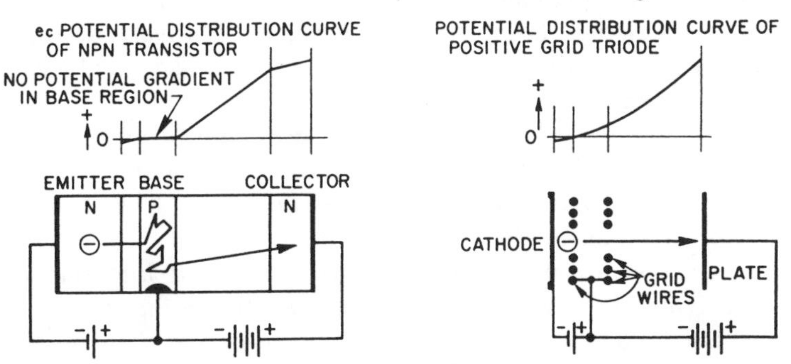

Fig. 3-13. A significant difference between transistor and tube action. The emitted charge carriers of the two devices encounter different field conditions in their respective control regions. In the transistor charge movement is by thermal diffusion. In the tube a potential gradient accelerates the charge through the grid region.

residual component of transistor collector current assumes an irreducible value governed almost entirely by temperature. Unfortunately, its rate of increase is very great with respect to increasing temperature.

SUMMARY

In the study of transistors we encounter two basic structures, two mobile charge carriers, two sources of creation of mobile charges, and two driving forces for mobile charges. We find two basic materials used, as well as two types of impurity materials. We shall see in a subsequent chapter that there are several diverse ways of forming the junction structure. Common to all structures is a functional dependency upon temperature. Finally, we shall have to enhance the agility and flexibility of our thinking by nimbly jumping from the characteristics attending any one of three basic circuit configurations to another. Most of these factors have been at least briefly discussed in this chapter. It would now be appropriate to resolve any confusion that may have arisen by summarizing the salient features of transistor action.

1. Currents in doped semiconductor substances are the net effect of two charge carriers, electrons and holes.

2. By controlling the relative doping in two adjacent conduction regions during manufacture, either electron or hole conduction can predominate.

3. A depletion layer, that is, a region relatively devoid of mobile carriers other than those thermally generated, exists between adjacent conduction regions.

4. The depletion layer acts as a space-charge region. For the most part, potential gradients in the transistor exist only across the depletion layers. In a sense, the depletion layers are the "vacuum" regions in the transistor.

5. Charges are not electrostatically accelerated into the base-emitter depletion layer. However, once they diffuse into this region under the influence of thermal agitation, they are then impelled by the force of the electric field.

6. In the complete transistor structure, charge carriers are injected from a forward-biased diode section into a reverse-biased diode section.

7. Power amplification ensues because virtually the same current is transferred from a section of low resistance, the emitter-base diode, to a section of high resistance, the base-collector diode.

8. The forward conduction bias applied to the emitter-base diode controls the charge injection into the reverse-biased base-collector diode. The circuitry function of the base region is analogous to that of the grid of a vacuum tube.

9. n-p-n and p-n-p structures involve similar concepts of transistor action, but the roles of electrons and holes are interchanged in the two arrangements. The n-p-n transistor is somewhat more analogous to the vacuum tube, in that electrons are the important charge carriers.

10. From physical considerations, it is natural to deal with the common-base transistor circuit, which is analogous to a grounded-grid vacuum-tube circuit. However, circuitry considerations lead us to consideration of the common emitter circuit which is analogous to the conventional grounded-cathode vacuum-tube circuit. In either case, the analogy between transistor and tube operation is enhanced by assuming that the tube operates in its positive grid region.

4. transistor structure and materials

INTRODUCTION

Purity of Semiconductor Material. A basic knowledge of the materials and processes employed in the manufacture of transistors provides an additional path of understanding of transistor action. Although there are different ways of producing the junction structures, they all are initially dependent upon intrinsic material of extreme purity. It is true that the required conduction characteristics are brought about by the presence of impurities. However, in order to attain the required control over the concentration of such impurities, it is necessary in the first place that the parent material be almost completely free from contamination.

Toward this objective, ingots may be made by melting small bits of powdered oxides of semiconductor material in a hydrogen atmosphere (see Fig. 4-1). The solid ingot obtained after cooling is then subjected to a "zone refining" process. In this process the ingot is placed in a graphite or quartz container and then slowly drawn through one or more high-frequency induction heating coils. The portions of the ingot exposed to the intense field are melted. As the ingot is pulled through the coils, the melted zones move progressively toward one end. In so doing, they sweep impurities into the last portion of the ingot exposed to the high-frequency field (see Fig. 4-2).

The purity of refined semiconductor material is determined from conductivity measurements. When the substance approaches its intrinsic state, it is known from its low conductivity that a negligible number of impurity atoms are present.

Growth of Large Single Crystals. The next step consists of growing a large crystal. The exact way in which the crystal is grown depends upon the type of junction transistor desired. One type of junction structure yields the so-called *grown junction transistor.* The crystal for such transistors is a large one from which a number of n-p-n or p-n-p junctions are sliced. To grow such a crystal, a tiny crystal is lowered into a molten mass of the substance, then slowly withdrawn at a controlled rate. A portion of the liquefied mass clings to the seed and solidifies. In turn, the solidified material draws up more of the liquid, which likewise solidifies and pulls up more melted material. In this way the crystal-lattice structure of the original "seed" is continuously supplemented by more and more material. Inasmuch as the atoms are arranged in the lattice structure in repetitive pattern, the adding of such atoms constitutes a growing process wherein the substance is not altered. We may say that the function of the seed is to provide an atomic blueprint for the formation of the large crystal.

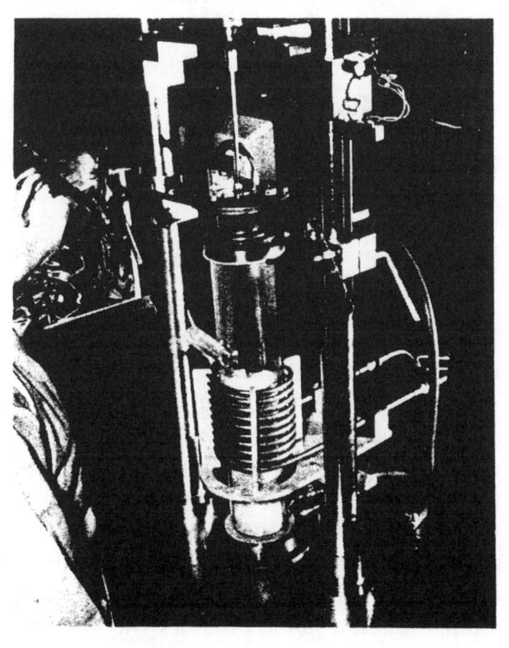

Fig. 4-1. Forming a silicon ingot. Ingot is slowly drawn upward from melt. Heat is supplied to melt from high-frequency induction generator via copper tubing coil. *(FairchildSemiconductor Corp.)*

FOUR BASIC MANUFACTURING PROCESSES

Although the fabrication techniques about to be described do not exhaust the number of structures based upon different manufacturing processes, they are representative of the four principal methods of making junction transistors. These can be classified as (1) grown, (2) alloy, (3) diffusion, and (4) electrode-position methods.

It is entirely possible that the end result of all of these techniques is essentially similar. However, some evidence exists wherefrom we may infer that the electrode-position method establishes a somewhat different kind of p-n junction from the junctions resulting from the other methods. The alloy and diffusion methods of producing junctions are not basically different; the term *alloy junction transistor* generally describes a transistor in which emitter

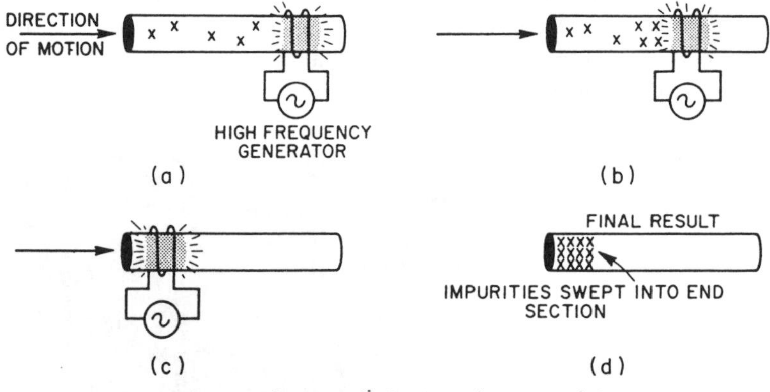

Fig. 4-2. Purification of semiconduction ingot by zone refining process.

and collector regions are formed by fusing materials of appropriate conductivity type to the body or surfaces of semiconductor material, which material then serves as the base region.

On the other hand, the diffused junction transistor is generally understood to be one in which the base region is obtained by fusing a small amount of material of appropriate conductivity type into a relatively large pellet or wafer of semiconductor material. An evaporation process, or another alloy-produced union, then establishes a collector-base junction. An ohmic contact made to the semiconductor wafer provides the emitter lead.

In the ensuing paragraphs, at least one transistor representing each of these four processes will be discussed. We shall also see that two or more of these processes can be combined to produce certain desired transistor characteristics.

The Grown Junction Transistor. In the grown junction transistor, the alternately doped regions are formed during crystal growth. This is accomplished by dropping donor (n impurity) and acceptor (p impurity) pellets into the large molten mass of parent material (see Fig. 4-3). If the pellet of n substance, such as arsenic,

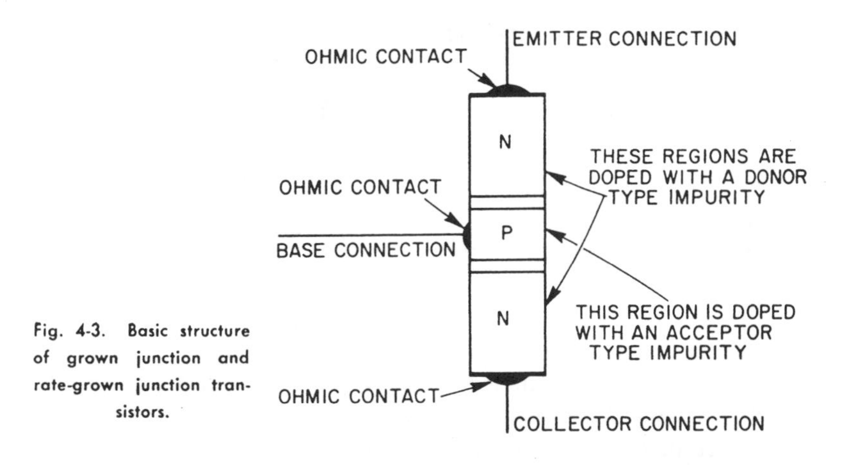

Fig. 4-3. Basic structure of grown junction and rate-grown junction transistors.

is dropped into the melt, this impurity rapidly diffuses throughout the melt and enters the body of the growing crystal.

Next, the doping of the crystal is reversed by feeding a sufficient amount of substance, such as gallium, to the melt. If the concentration of such p material exceeds that of the previously introduced n material, p-type doping then overcomes *n* doping. As a result, a p region now forms adjacent to the previously formed n region in the growing crystal.

The third step involves the injection into the melt of sufficient n material to overcome the p-type conductivity. This results in the formation of the final n doped region in the growing crystal. We see that we have grown a single large crystal from which it is possible to slice a number of n-p-n structures. A similar procedure is employed for obtaining grown junction p-n-p transistors. Fundamentally the same process is used in making silicon transistors.

THE RATE-GROWN JUNCTION TRANSISTOR. Another junction-forming technique is used to produce "rate-grown junction transistor." The procedure is somewhat similar to that employed for making grown junction transistors (see Fig. 4-3). However, donor and acceptor impurities are *simultaneously* present in the melt. The selection of one or the other for entry into the growing crystal is governed by the *rate of withdrawal* from the melt. For slow rates

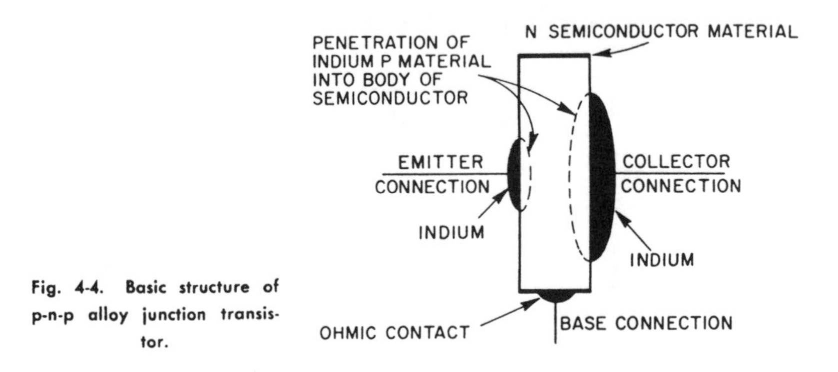

Fig. 4-4. Basic structure of p-n-p alloy junction transistor.

of withdrawal, n material enters the crystal, and the p material remains in the molten state. Faster withdrawal rates are favorable for the pulling out of p dominant material. This selective process is, in part, due to the fact that the n impurities have lower melting temperatures than the p impurities. Thus by alternating crystal withdrawal rates, alternate doped regions can be progressively formed in the growing crystal. This process is generally repeated several times, so that a large number of junction structures may be sliced from the grown crystal.

The Alloy Junction Transistor. The alloy junction transistor is made quite differently from the two grown junction types. The process will be described for p-n-p germanium structures. Small wafers are sliced from the body of an n-type crystal. Pellets of indium are placed on the two surfaces of the wafers, and the resultant assembly is heated in a hydrogen atmosphere. The indium melts and with the germanium forms a chemical and mechanical bond similar to that produced in soldering. The indium atoms penetrate into the body of the germanium material wherein they behave as acceptor, or p-type impurity atoms. One of the indium pellets placed on the wafer is larger than the other. This pellet becomes the collector electrode, whereas the smaller pellet becomes the emitter electrode (see Fig. 4-4).

N-p-n alloy junction transistors are made in a similar way, but the process makes use of p-type material to which is alloyed a material whose atoms behave as donors within the body of the parent substance. The alloy junction structure is particularly well suited for transistors designed to handle high power. In such transistors, the collector button is welded to a heavy copper plate, which in turn, through the mounting provision, establishes good thermal contact with the chassis or "heat sink." In this way the temperature of the base-collector junction is maintained within operational limits for relatively high values of power dissipation.

Diffused-Base Transistor. A modified alloying process is used in making the diffused-base or "mesa" transistor (see Fig. 4-5). A small pellet of n-type material is placed upon a surface of a p-type germanium wafer. This assembly is placed in an oven whereupon a thin layer of n impurity atoms diffuse into the surface of the germanium. Surplus n material is then etched away. A second diffusion process unites a film of p material such as aluminum with the etched surface of n material. This p material alloys with the n material and produces a p-n junction. A broad ohmic contact is made to the original germanium to form the collector electrode. The diffused base transistor, like the surface barrier transistor, has very good high-frequency possibilities because the base region can be made extremely thin.

The Intrinsic Barrier (n-p-i-n or p-n-i-p) Transistor. One of the ways in which an extension of high-frequency transistor action can be achieved is to lower the effective capacitance of the base-collector junction. Operation of transistors at relatively high collector voltage helps in this respect because the depletion layer between base and collector regions widens, thereby dcreasing the

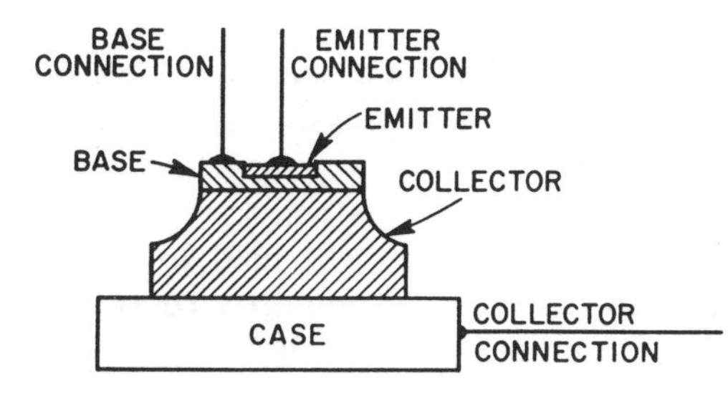

Fig. 4-5. Basic structure of diffused-base transistor.

collector capacitance. This operational technique has its limits however. If we increase the collector voltage too far, transistor action is lost due to the occurrence of avalanche in the base-collector diode, or to another phenomenon known as *punch-through.*

Punch-through involves the widening of the base-collector depletion layer to the extent that it extends entirely across the base region and virtually joins the emitter-base depletion layer. Collector capacitance can be decreased, and the punch-through voltage raised by the insertion of an artificial depletion layer

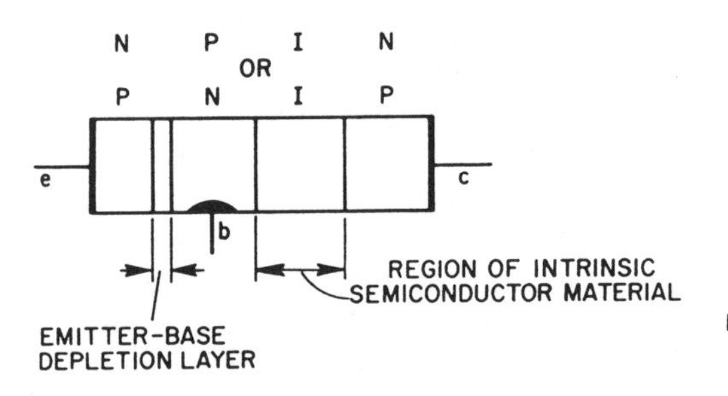

Fig. 4-6. The intrinsic-barrier transistor.

between base and collector regions. Such an artificial depletion layer consists of a region of intrinsic material (see Fig. 4-6).

Such a transistor structure also improves high-frequency operation by allowing heavier doping of the base region than would otherwise be possible. This results in low resistance in the base region, a required parameter for high-frequency operation. Since the intrinsic layer provides the base-collector junction with a nearly fixed width depletion layer, the base can be made very thin without danger of punch-through at moderate voltages. A thin base region enhances high-frequency operation inasmuch as less time is then needed for the injected charge carriers to diffuse across it. Such an improvement is analogous to decreasing transit time in a high-frequency vacuum tube.

The "Drift" or Graded Junction Transistor. Similar objectives to those attained in the n-p-i-n and p-n-i-p transistors can be realized in a somewhat related structure. If the doping of the base-layer gradually decreases in traversing from its emitter to its collector side, the base-collector depletion layer will tend to extend less and less into the base region as increased impurity concentration is reached. This mitigates against punch-through.

An additional desirable effect results from this structure. For other junction transistors, it has been emphasized that the charge carriers traverse the base region under influence of a concentration gradient produced by thermal agitation. No appreciable electric field impels these carriers. If somehow these carriers could be subjected to an electric field while in the base region, transit time could be reduced, with attendant improvement in high-frequency performance. It so happens that if the impurity doping of a semiconductor region is more or less *gradually* stepped from a high to a low concentration, a voltage gradient will exist along the tapered impurity region. This voltage gradient is our "built-in" internal junction voltage in modified form. Inasmuch as our "junction" is no longer abrupt, it follows that the internal "junction" voltage is not abrupt either. In a sense, we have merged much of the base region with the base-collector depletion region (see Fig. 4-7). One is not distinguishable from the other over a substantial portion of the material in the vicinity of the base electrode.

If the decrease in impurity concentration takes place in the direction proceeding from emitter to collector, charge carriers

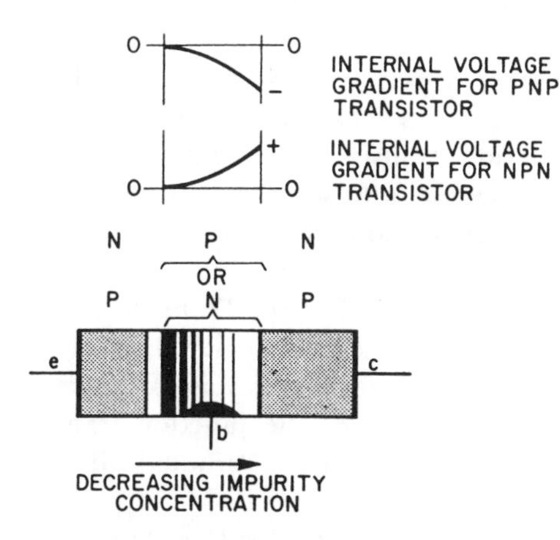

Fig. 4-7. The graded junction or "drift" transistor. Due to graded doping in base region, the base-collector "depletion layer" has no abrupt boundary in the base region.

injected from the emitter region will be speeded in their transit through the base region by this voltage gradient. Although the voltage gradient prevailing in the base region is somewhat modified by the application of bias voltages, we note that it is an inherent feature of the drift transistor and is not primarily dependent upon bias potentials.

The Surface Barrier Transistor. Yet another junction-forming technique is used in making surface barrier transistors (see Fig. 4-8). Both surfaces of a wafer cut from n-type germanium are subjected to jets of an electrolyte. The electrolyte in a typical instance is a solution of an indium salt. D-c currents are passed through the germanium and the jets in such a direction as to *etch* the

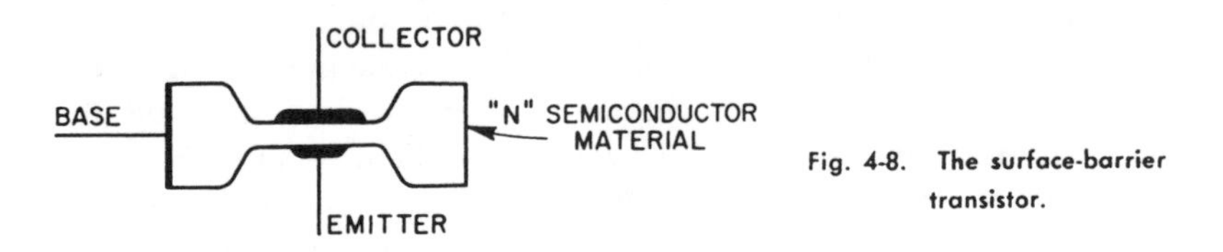

Fig. 4-8. The surface-barrier transistor.

surface of the wafer. When sufficient material has been removed from the wafer, the directions of the currents are suddenly reversed. Indium metal is now *plated* on the two surfaces of the germanium wafer. Abrupt junctions are thereby formed at the surfaces of the etched germanium.

This structure can be more precisely controlled than one produced by the alloy junction technique. Very thin n-type base regions are produced, and such transistors yield excellent high-frequency performance. The formation of the alternate conduction regions is somewhat different in this structure than in other junction transistors, inasmuch as heat is not applied to cause appreciable diffusion of indium into germanium. This will be elaborated upon in our subsequent discussion of the point-contact transistor.

The Point-Contact Transistor. The proper scope of this book is to convey basic understanding of *junction* transistor action. It would, however, be relevant to make a few statements concerning the predecessor of the junction transistor, the *point-contact transistor* (See Fig. 4-9). Paradoxically, although this was the first semiconductor triode capable of producing amplification, its action is not so well understood as that of the later and more successful junction transistor.

A p-n-p point-contact transistor is made by pressing two needle-like metallic points,separated by a few thousands of an inch, against a surface of a wafer of n-type material. One point constitutes the emitter, the other serves as collector. A large-area metallic contact is established at the underside of the wafer to provide the base connection. Circuit-biasing is similar to that employed in junction transistors. Apparently reversals in conduction properties exist between the metallic points and the semiconductor material when forward and reverse currents flow. The emitter point is often welded to the germanium wafer in order to stabilize the characteristics of the structure. In a sense, the collector point is also welded to the germanium but by a unique method and for a different reason. It is found that the characteristics of the point-contact transistor may be controlled during manufacture by pulsing a current through the collector point.

A theory of operation which affords some insight into point-contact operation involves a peculiar surface phenomenon of semiconductor material. Regardless of the conduction characteristics of the bulk material, there appears to be a tendency for the surfaces to assume the charge state of p-type material. Furthermore, the

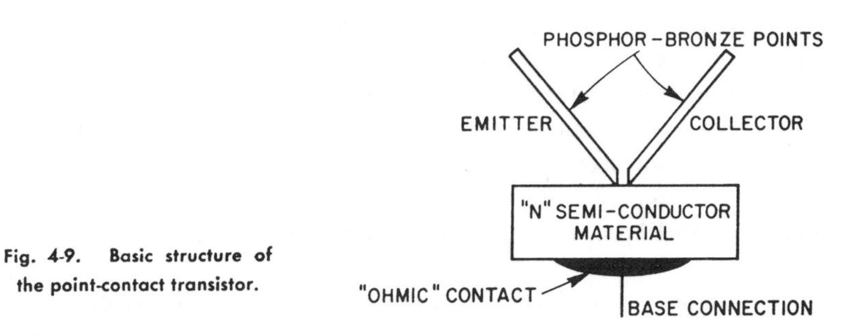

Fig. 4-9. Basic structure of the point-contact transistor.

internal bulk of a slab of intrinsic germanium actually displays n-type conduction characteristics. This is due to the fact that, despite the equality of thermally generated electron-hole pairs, the electrons have a higher mobility than the holes, that is, their rate of movement is greater. The p-type surface in conjunction with the n-type interior provides us with a natural arrangement for forming transistors. In essence, we do this by introducing electrodes at the surfaces.

There are many diverse ramifications to this theory; as presented in its simple form, however, it offers a basic explanation of the point-contact as well as the surface-barrier transistor.

In the light of this theory it would appear that, to a first approximation at least, the essential difference between the surface-barrier and the point-contact transistor is one of geometry. A peculiarity of point-contact transistors is that their current amplification factor α is greater than unity. In some manner, carriers are multiplied in the vicinity of the collector point. (A similar end result is possible in the junction transistor as well. For example, excessive heat or excessive reverse voltage applied to the collector junction can increase generation of electron-hole pairs, so that a current-multiplying mechanism operates to produce α greater than 1.) In a vacuum tube, a similar end result could conceivably result due to secondary emission from a positive screen grid. In such a case, the plate current would exceed the cathode current, yielding α greater than 1. The grounded, or common-grid circuit serves the best interest of illustration by analogy in this case.

CONTACTS

Ohmic vs. Junction Contacts. There may be a tendency towards confusion in distinguishing so-called "ohmic" contacts from electrodes which, by intention, participate directly in transistor action. The difference is one of material and of geometry. If the contact is intended as a medium for conveying current to or from semiconductor material, but is not intended as part of a junction region, the desired connection will be *ohmic.* Such a connection ideally serves the same purpose as an ordinary soldered splice.

If the contact is intended to function as a p-n junction, material of opposite-type conductivity to the contacted semiconductor is used. For example, indium can contribute donors to semiconductor material via physical diffusion. If contacts are sharp points, the field intensity resulting from biasing current, suffices to dislodge charge carriers in contacted n-type material. This produces the effect of tiny p-n junctions and is profitably exploited in the point-contact transistor. The simplified theory presented is reinforced by the fact that transistor action is improved by using points containing a p-type substance, such as phosphorbronze, for example. In the latter case, the phosphor apparently acts as an acceptor, that is, a source of holes for a p-n-p structure. A sharp point never provides an ohmic contact to the body of semiconductor material.

Difference is One of Degree. Actually the difference between ohmic and non-ohmic contact is a matter of degree. When unlike materials are in physical contact, it is difficult to produce linear or "ohmic" electrical conductivity, because a "weak" p-n junction generally is produced by the different energy gaps of the materials. The use of the descriptive term "ohmic" with respect to transistor connections therefore has a relative connotation. It infers that charge-injection and depletion-layer phenomena are negligible compared to what occurs within the transistor structure proper. In general, ohmic contacts are obtained by using a connecting material of the *same* conductivity type as the semiconductor region to which such contact is desired. Thus, indium diffused into the surface of p semiconductor material establishes a contact which is essentially "ohmic." Indium is a trivalent element and acts as an acceptor or p substance when diffused into germanium. Indium diffused into the surface of *n-type* germanium produces a p-n junction. In ohmic contacts the voltage drop across the contact elements is, ideally, *directly proportionate,* to current flow. In p-n junctions the voltage drop is an *exponential function* of current flow.

SEMICONDUCTOR MATERIALS

Germanium vs. Silicon. As may be said about people, germanium and silicon are similar; but it is the little differences which are important. Both of these substances are included in group IV of the periodic table of chemical elements. Neither can be said to display properties which permit a clear-cut designation as either metallic or nonmetallic. Rather, they are both transitional elements. They are capable of displaying both metallic and nonmetallic properties; at the same time they can behave in a manner mid-way between metals and nonmetals.

Of particular relevance to transistors, germanium and silicon are neither insulators, as are the nonmetals, nor are they conductors as are the metals. At room temperature, these two elements are just beginning to make the transition from insulators to conductors. Conduction becomes possible, we recall, when outer orbit electrons are freed from the binding force of the atom to move through a material as elemental carriers of the electric current. In germanium and silicon, the application of a weak electric field suffices to free large numbers of electrons for conduction. (Even less field force is required after addition of donor or acceptor impurities.) It is here that we find ourselves face to face with the important difference between the two semiconductor materials.

Temperature Characteristics of Germanium and Silicon. For a given temperature, germanium has a less firm hold on its outer-orbit electrons than has silicon. The measure of this property is given by the energy gap expressed in electron volts. We say that germanium has a *lower* energy gap than silicon. We might have anticipated this from the fact that the four outer-orbit electrons of the germanium atom are in the fourth outermost ring from the nucleus, whereas the corresponding electrons of the silicon atom occupy the third outermost ring. A reverse-biased p-n junction, such as we have in the base-collector diode of a transistor, permits a small saturation current to flow because of thermally generated minority carriers. Higher temperature supplies more thermal energy which, in turn, increases the saturation current. As the temperature is increased, a point will be reached wherein the current due to such minority carriers is of the same order of magnitude, or even exceeds, the current which would be provided by majority carriers under the condition of forward conduction. It is obvious that the forward-to-backward conduction ratio, that is the "goodness" of a p-n rectifier suffers degradation with rising temperature. Above room temperature, germanium will always be worse in this respect than silicon. In a transistor, we begin to lose transistor action as the collector saturation current becomes an appreciable fraction of the emitter current. Under such a condition the injected carriers from the emitter no longer maintain control of collector current. Indeed, a collector junction having attained an elevated temperature, tends to produce increased collector current, which in turn heats the junction further.

We see that the cycle of events is cumulative, and can even be destructive. This condition is aptly described as *thermal runaway.* The effect of the alternately doped regions is largely overcome, and the electrical conduction approaches that of intrinsic material which, though negligible at room temperatures, becomes many orders greater at elevated temperatures.

Temperature Effects on Conduction. The temperature sensitivity of conduction is of paramount importance in adapting an intrinsic semiconductor material to a practical transistor application. It is in this regard relevant to consider the formulas so describing germanium and silicon. We let C_g represent the conductivity of germanium, and T the absolute temperature. From conductivity measurements at different temperatures we then obtain the relationship:

$$C_g = 4.3 \times 10^4 e^{-4350/T} \text{ ohms/cm}^2$$

Similarly, if C_s is the conductivity of silicon, we have the relationship

$$C_s = 3.4 \times 10^4 e^{-6950/T} \text{ ohms/cm}^2$$

INTERPRETATION OF FORMULAS. The numbers and their arrangement in these two relationships tell us that, for a given temperature, the conductivity of intrinsic germanium exceeds that of intrinsic silicon. Furthermore, the rate of increase in conductivity with rising temperature is greater in germanium than in silicon. It is the latter factor which imposes the main restriction to high-temperature operation of germanium. This is because in a transistor the reverse-biased base-collector diode establishes a depletion layer which behaves as intrinsic material.

We can therefore infer from these formulas that the reverse current in a germanium transistor must become excessive at a lower temperature than in the case of a silicon transistor.

Forward and Reverse Conduction. Because we must describe first one, then the other, it might appear that forward and reverse conduction represent qualitatively different phenomena. That the contrary is actually the case, is revealed by a single formula which equates current in terms of applied voltage for either type of conduction. At room temperature, the current I passed through a p-n junction diode is given by the relationship

$$I = I_s (e^{40V} - 1)$$

where I_s is the saturation current, that is, the current resulting from the application of reverse voltage; and where V is the applied voltage.

Mathematical analysis of this equation shows that forward and reverse conduction simply represent operation under two different voltage conditions. The quantity $(e^{40V} - 1)$ is substantially -1 for any value of $-V$ numerically greater than -0.1 volt. On the other hand, when we apply $+V$, corresponding to forward conduction, the quantity $(e^{40V} - 1)$ grows very rapidly. This state of affairs is graphically illustrated in Fig. 4-10. This formula applies both to germanium and silicon inasmuch as it indicates only relative current in terms of I_s.

I_s for germanium at room temperature is much greater than for silicon. Therefore a given forward-conduction voltage will produce a heavier forward-conduction current than for silicon.

The current in a p-n junction increases exponentially with temperature. This causes the room temperature difference in reverse current between germanium junction diodes and those made of silicon to diverge increasingly as temperature is raised. When

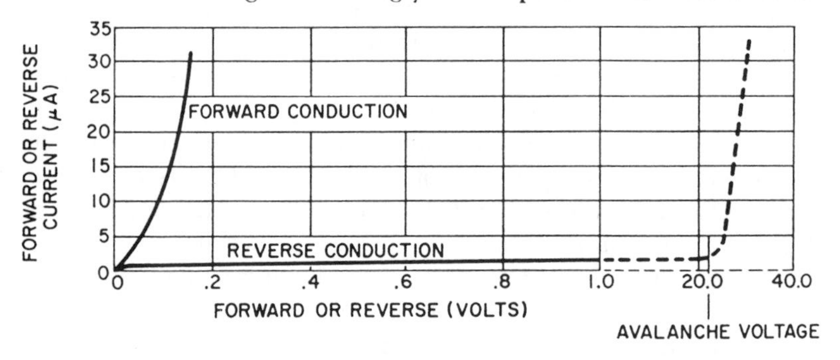

Fig. 4-10. Conduction in typical germanium junction diode.

collector junction temperatures approach the vicinity of 90°C, the collector current existing without emitter bias (I_{CO}) is often no longer low enough for good transistor action in germanium units. On the other hand, silicon transistors can be designed for operating temperatures in excess of 150°C.

High-frequency Considerations in Germanium and Silicon. The depletion layer in a junction diode behaves in the manner of intrinsic material. Not only can we deal with germanium and silicon as semiconductors, but we can also consider them as "semi-insulators." Accordingly the depletion layer constitutes a somewhat leaky dielectric. We can appreciate that the dielectric constant of the material must bear an effect upon high-frequency operation. Silicon is more favorably disposed in this respect, having a dielectric constant of 12 contrasted to 16 for germanium. This enables silicon transistors, other factors being equal, to be made with less collector capacitance, an important point for high-frequency performance. There are other frequency-determining parameters, such as fabrication geometry, base resistance, and carrier mobility. The latter item is much higher in germanium. Theoretically, the greater carrier mobility in germanium, makes this material appear to be better for operation at extremely high frequencies. Practically speaking, neither material can be said to have an overwhelming advantage for high-frequency operation, and the material used for the highest frequency transistors appears to depend upon the state of the art.

Other Operational Factors. The heat conductivity of silicon is somewhat better than in germanium. This, in conjunction with the higher energy gap of silicon establishes this material as very well suited for large power transistors. Yet another desirable feature of silicon for use in power transistors is the "self-healing" nature of excessive reverse voltage. The silicon junction withstands operation in its avalanche region much better than does germanium. In the latter material, the avalanche phenomenon tends to be more destructive. However, the silicon power transistors are considerably more expensive than the germanium power transistors and are therefore not so frequently encountered.

5. biasing the transistor

THE NEED FOR BIASING

Basically, similar objectives are served in biasing the transistor as are provided by the polarizing voltages in vacuum-tube operation. We shall see, however, that the transistor is somewhat the more critical of the two devices. This is primarily due to the residual collector reverse current I_{CO}, which can appear in the input circuit and emerge sufficiently amplified to drastically move a chosen collector operating point. The fact that I_{CO} is extremely temperature dependent, of course only serves to agitate the problem. Indeed, a cumulative sequence, known as *thermal runaway* is often destructive to the transistor.

TWO-BATTERY BIAS

The most direct, if not the simplest biasing arrangement makes use of two batteries, or other current sources. These are connected to provide suitable biases to the emitter-base diode and to the base-collector diode. The emitter-base diode is polarized for forward conduction; the base-collector diode is polarized for reverse conduction. The resultant arrangement would be classified as a common-base circuit for direct current. We note that the base connection is *common* to the two bias sources. This circuit is a fundamental one for approaching basic principles of both d-c and a-c operation of transistors (see Fig. 5-1). In this circuit the emitter current is slightly greater than the collector current. The difference between the two currents flows in the base lead. If we open the connection to the emitter lead, we find that the residual current I_{CO} persists in the collector circuit. The described circuit conditions constitute the definition of I_{CO}, for it is the residual collector current in the common-base circuit when the emitter is open. We note that we are dealing with the base-collector diode as an isolated diode; that is, no transistor action is involved. As previously pointed out, this reverse current is practically independent of voltage, but increases very rapidly with temperature.

Component Parts of Collector Current. From the foregoing discussion, it is evident that the total collector current in the two-battery common-base bias arrangement is made up of two component parts. One of these is the residual or saturation current

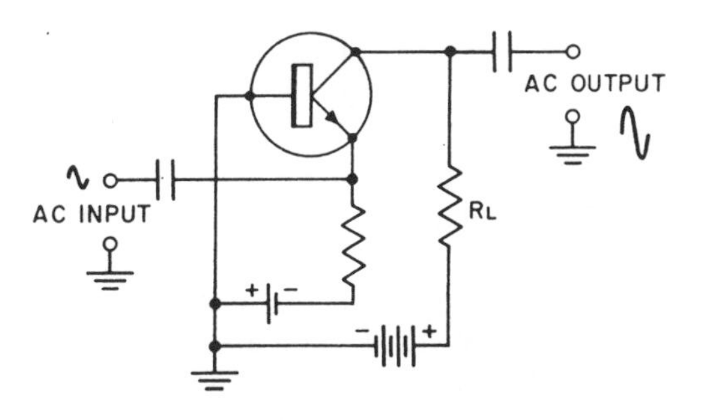

Fig. 5-1. Two-battery bias arrangement in d-c common-base circuit.

I_{CO}. The other is the portion resulting from carriers injected from the emitter through the base, and into the collector region, that is, from transistor action. If this portion of collector current is much larger than I_{CO}, the effect of I_{CO} at room temperature may not be of consequence. However, at elevated collector junction temperature, I_{CO} will be many times its value at room temperature. Temperature rise may be developed by the collector junction either from increased ambient temperature or from junction power dissipation.

The portion of collector current resulting from carrier injection will, on the other hand, remain of the same order of magnitude for wide temperature shifts. Therefore, an inconsequential I_{CO} at ordinary room temperature might, under the influence of additional heat, increase sufficiently to displace the collector-current operating point. Other things being equal, silicon transistors have much better bias stability than germanium transistors because I_{CO} is always much lower at a given temperature. Whenever possible, transistors should be biased to a sufficient collector current to exceed I_{CO} by a factor of at least 10 under the most adverse operating conditions.

Common-Emitter Bias Circuit. While discussing two-battery bias supplies, it would be instructive to change the battery connections so that we have a common-emitter circuit for direct current (see Fig. 5-2). This is a much less desirable biasing arrange-

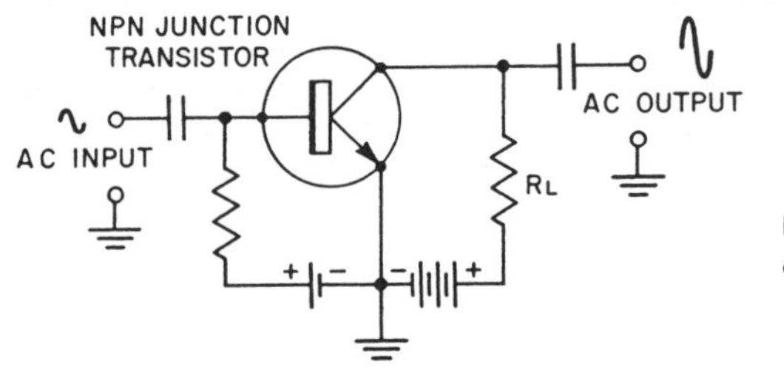

Fig. 5-2. Two-battery bias arrangement in d-c common-emitter circuit.

ment. This evaluation holds true notwithstanding that either of the arrangements can operate common emitter or common base for ac.

Let us investigate the behavior of this biasing arrangement. Due to the physical structure of the transistor, collector current completes its circuit through the emitter-base diode. In so doing it adds to the number of carriers injected into the base region. The over-all result of this is that I_{CO} is amplified by a factor of $1/(1-\alpha)$ which is very close to β (beta), the current amplification factor of the grounded emitter circuit.

$$\beta = \frac{\alpha}{1-\alpha}$$

If we consider a typical α of 0.95, it is evident that the residual collector current in this circuit is 20 times greater than in the d-c common-base circuit. A temperature rise will cause the current in the common-emitter circuit to exceed I_{CO} in the common-base circuit by many times 20 (see Fig. 5-3). A further practical disadvantage of this arrangement is that due to the polarity requirement of the two transistor diodes, two separate batteries rather than a tapped battery must be used.

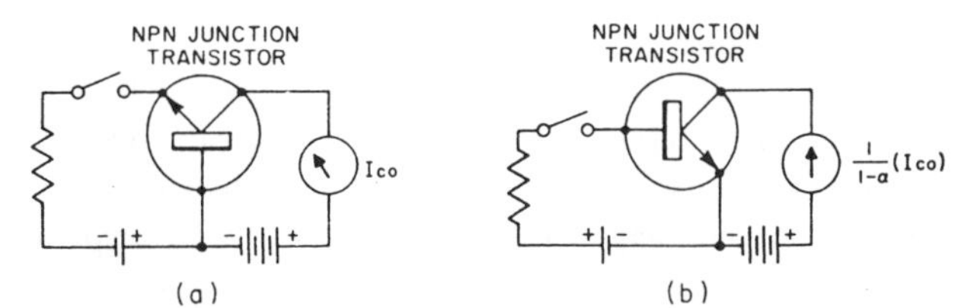

Fig. 5-3. The collector current with no forward bias applied to input diode: (A)collector battery is connected from collector to base, producing common-base circuit for dc; (B) collector battery is connected from collector to emitter, producing common-emitter circuit for dc.

Stabilization by Emitter Resistance. Despite the relative instability of the d-c common-emitter bias circuit, it would be expedient to devote further attention to its properties. This is because the d-c common-emitter circuit lends itself readily to single-battery operation through a modification. No battery taps are then required, and the over-all convenience exceeds other arrangements. Fortunately a rather simple technique confers considerable stability to the common-emitter circuit. This consists of the insertion of resistance in series with the emitter lead (see Fig. 5-4).

Such a resistance accomplishes two desirable things. First, the emitter-base diode should be energized from a constant-current source. Otherwise temperature and operational changes will affect the injection of mobile carriers and the effect will be amplified in the collector circuit. An emitter resistance results in substantially constant emitter-base current if it is comparable to—or preferably several times larger than—the effective internal emitter-base resistance. Secondly, such an emitter resistance enhances stability by virtue of its unique position in the transistor circuit. This resistance provides a common path for input diode and output diode currents. It thereby constitutes a means of producing feedback. The feed-

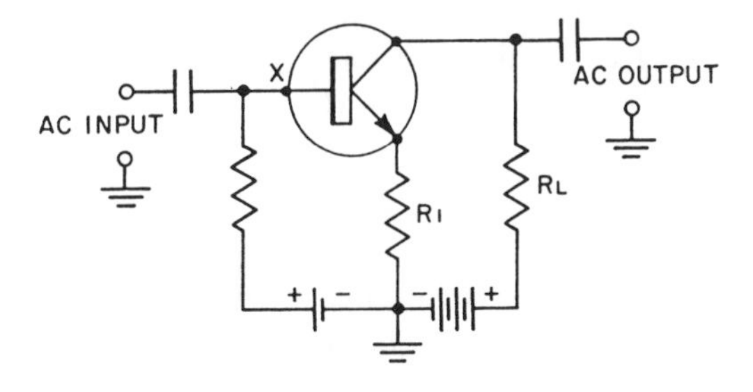

Fig. 5-4. Simple two-battery stabilized bias arrangement for d-c common-emitter circuit.

back is negative because an increase in collector current produces an increased voltage drop across this resistor, which changes emitter-base bias in such a direction as to oppose the increase in collector current. Thus, the actual β of the transistor is no longer effective as the current gain factor in the circuit. By bypassing this resistance with an appropriate capacitor, the transistor β is readily recaptured for a-c operation.

Effect of External Collector Resistance. In practical circuits the collector circuit generally contains either a resistance or a transformer winding. However the collector current is, by the na-

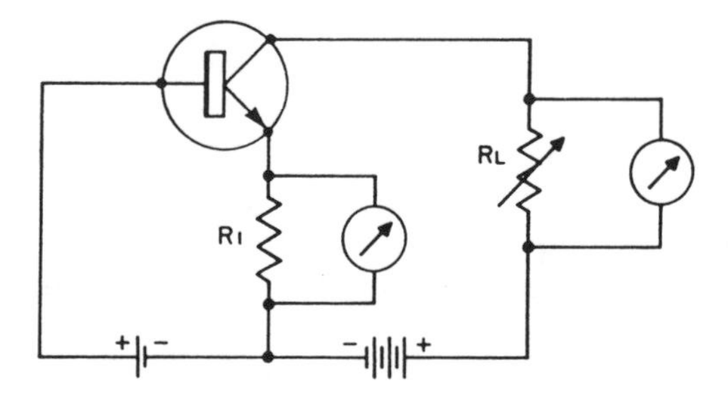

Fig. 5-5. Relative effects on d-c feedback stabilization produced by emitter and collector resistances.

ture ofthe transistor substantially independent of applied voltage. For this reason, external collector resistance does not appreciably add to general bias stability. Indeed, the stabilizing effect of a given emitter resistance is reduced by resistance inserted in the collector lead. (The voltage drop developed across such a resistance is not available for negative feedback.)

Note in Fig. 5-5 that the d-c voltage drop across R_1 produces d-c negative feedback which reduces the effective d-c amplification factor of the circuit. The higher this voltage drop with respect to the sum of both voltage drops (across R_1 and R_L), the greater the d-c stabilizing action. The d-c voltage drop across R_L is a lost voltage insofar as feedback stabilizing action is concerned. The effective feedback voltage due to collector current is directly proportional to $R_1/(R_1 + R_L)$.

SINGLE-BATTERY BIAS

The aforementioned considerations of bias requirements lead naturally to a single-battery bias circuit with emitter resistance.

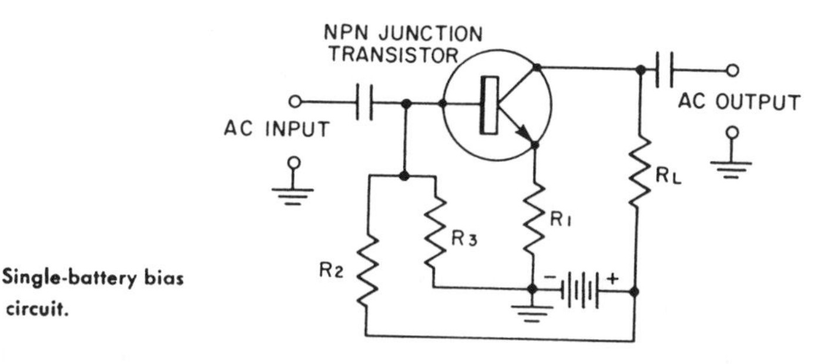

Fig. 5-6. Single-battery bias circuit.

The two required biases are obtained from a divider network connected across the battery. This circuit is shown in Fig. 5-6. The basic configuration is the d-c common-emitter arrangement. Despite the inherent bias instability of the simple common-emitter circuit, this modified circuit is capable of extreme stability when the resistances are properly chosen. The reason it is necessary to design *around* the common emitter circuit is that we encounter an embarrassing design situation when we attempt to provide the requisite biases for a common-base circuit from one untapped battery and a divider network. Whereas polarity requirements of the common emitter circuit are readily met by a single untapped battery and a divider, such is not the case in the common-base circuit.

Figure 5-7 shows the situation for a common-base circuit using an n-p-n transistor. We see that the required bias for the emitter is negative with respect to the base and that such a bias cannot be provided by the divider connected in conjunction with the battery. Because of these circumstances, the common-emitter arrangement of Fig. 5-6 has become the basic bias circuit—"basic" in that most other connections can be considered as simplifications or modifications thereof. We appreciate, of course, that each resistance dissipates power and the inherent efficiency of the transistor

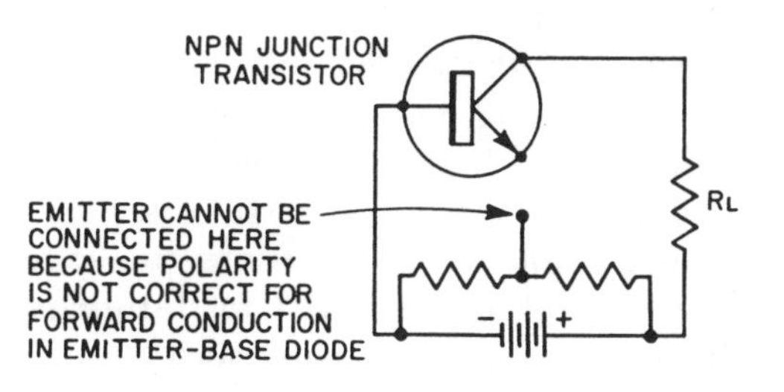

Fig. 5-7. Circuit showing that a divider network in conjunction with a single battery cannot provide both biases in d-c common-base circuit.

is not realized by virtue of the biasing network. However, if the resistances are wisely chosen, this will not be a significant factor in most practical applications. It still remains true that the power consumption of the transistor and its associated circuitry is a small fraction of that required by tubes to accomplish similar ends.

THE STABILITY CRITERION, S

In order to make our discussion of bias stability meaningful, we introduce the stability criterion S. S is the number of times change in I_{CO} appears magnified in the total collector current. In the simple, unstabilized common-emitter circuit, S is approximately equal to the common-emitter current-gain factor β. Inasmuch as β may range typically between 20 and 100, an S value of, say, 2 or 3 represents a significant improvement in bias stability. In any event the lower the value of S, the greater the stability. The approximate value of S is readily obtained from the formula

$$S = \frac{1 + (R_1/R_2) + (R_1/R_3)}{1 - \alpha + (R_1/R_2) + (R_1/R_3)}$$

If we let $R_1 = 0$ in this formula, we find that $S = 1/(1 - \alpha)$ which is close to the value of β $[\beta = \alpha/(1 - \alpha)]$. If α is not known, a good approximation to S may be obtained in practice by assuming α to be 0.970.

Elimination of R_3 in Modified Bias Circuit. We note that the load resistor R_L does not appear in this relationship. We cannot stabilize collector current by means of resistance inserted in the collector lead because collector current is not governed primarily by collector voltage, but rather by that input diode of the transistor. The resistors associated with the input diode are R_1, R_2 and R_3 (see Fig. 5-6). This formula further indicates that stabilization is possible in a circuit in which R_2 is not present, that is, where R_3 is infinite. Consequently, the simpler stabilized circuit contains only R_1 and R_2 (see Fig. 5-8).

An advantage of dispensing with R_3 is that a source of battery drain is thus eliminated. However, it often happens that R_2 must then be in the neighborhood of several or more megohms. Such a high resistance is sometimes undesirable in this application because its value is subject to disturbance by moisture and dirt. One's fingers placed across such a high resistance could cause destruction of the transistor. On the other hand, there is less likelihood that the input signal source will be unduly loaded by this single high resistance.

Simplified Calculation of Resistances Associated with Bias Network. The values R_1, R_2, and R_3 can be calculated from three simplified relationships. The results are approximate but sufficiently accurate for a wide variety of practical circuits. In the following formulas, the initial assumption is made that I_{CO} is small compared to total collector current I_c. In this way, I_{CO} need not be considered as a necessary factor in the formulas. This assumption may not always be true, but it is generally justified in many practical circuit applications.

$$R_1 = \frac{E_b - E_c - (I_c R_L)}{I_c}$$

$$R_2 = \frac{S - 1}{I_c/[E_b - E_c - (I_c R_L)] - I_c/E_b}$$

$$R_3 = \frac{E_b(S - 1)}{I_c}$$

where R_L = load resistance
E_b = battery voltage
E_c = collector voltage
I_c = collector current
S = stability factor

The relationships given by these formulas become increasingly accurate as I_c exceeds I_{CO}.

General Procedure in Design of Bias Network. We may begin by selecting an S value of approximately 0.1 of the β of the transistor. In any event, an S of say, 3, should confer more than sufficient

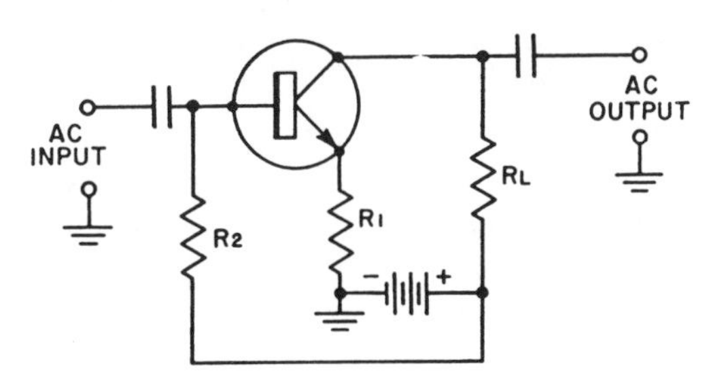

Fig. 5-8. Simplified single-battery stabilized bias circuit.

stability in most instances. From considerations of economics, practical application, and required voltage swing, we select our battery or supply voltage. R_L is governed by the a-c gain we wish to obtain. In the a-c common-base amplifier, usable gain commences for load resistances of several hundred ohms. In the a-c common-emitter amplifier, R_L values between 50 ohms and 50 kilohms are found in many practical circuits. Here again, the gain increases as R_L is raised. I_c is selected to harmonize with conditions imposed by R_L and E_b. I_c must be low enough to permit a suitable collector voltage E_c. The main requirement for E_c is that it is high enough to permit the amplified signal voltage to undergo its peak-to-peak excursions. Providing that battery economics does not interfere, it is generally best to make I_c nearly as high as signal-voltage swing permits. In this way we tend to make the ratio of I_c *to* I_{CO} as high as possible. This is a desirable operational feature as well as a condition for optimum calculation accuracy with the simplified formulas. When it is necessary to operate germanium transistors at low collector currents, it is often most expedient to calculate R_1 and R_3 but to determine R_2 experimentally. In so doing, one should start with a very high resistance, about several megohms, and progressively reduce the value until desired collector current is obtained. For small "audio-type" germanium transistors, the experimental procedure for determination of R_2 should be employed when the desired collector current is less than, say, 0.5 ma.

A Variation of the Basic Bias Network. We sometimes encounter a variation of the basic bias network as shown in Fig. 5-9. In this circuit R_2 acts as a negative-feedback path as well as an element of a voltage-divider network. In the event of an increase in d-c collector current, the collector voltage will be reduced by virtue of the increased potential drop developed across load resistance R_L. This, in turn results in a reduction in forward-biasing current applied through R_3 to the base. Decreased base current results in a reduction of collector current. The converse sequence of events occurs in the event of a tendency for collector current to decrease.

We see that d-c negative feedback such as provided by R_2 in this connection helps stabilize the collector current operating point. This arrangement is difficult to design or adapt, because R_2 also functions as an a-c feedback path and thereby reduces a-c gain.

It is not possible to bypass R_2 as in the case of R_1. Indeed, a capacitor connected across R_2 will only further reduce a-c gain. Generally, in order to obtain a maximum ratio of d-c to a-c affect, R_2 should be high. This in turn may require a higher voltage from the bias supply than would otherwise be the case. A practical approach to the design of this network is to first obtain operation from the conventional network with R_2 connected to the battery side of R_L. R_2 is then disconnected and a new value of R_2 is connected from base to collector in compliance with desired collector current.

OTHER BIASING CONSIDERATIONS

Relevant to Transistor Type. Everything said concerning bias circuits applies in like manner to both n-p-n and p-n-p transistors, except for battery polarities. If we are constructing a circuit in which an n-p-n transistor is specified, but we wish to use p-n-p

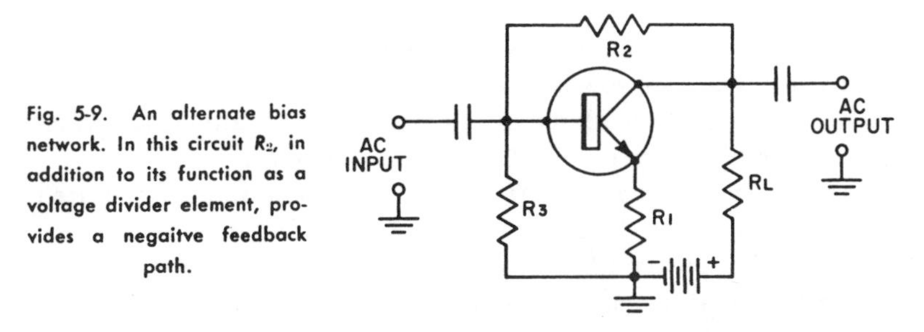

Fig. 5-9. An alternate bias network. In this circuit R_2, in addition to its function as a voltage divider element, provides a negaitve feedback path.

type having similar characteristics, it is only necessary to reverse the polarity of our bias supply. Because of their relatively low I_{CO}, silicon transistors often require, other things being equal, less attention in the matter of bias stabilization. Germanium and silicon transistors which appear to be very similar are not always interchangeable in the same bias network. This is because the emitter-base diode of the silicon transistor requires a higher voltage to produce a given emitter-base current than in the germanium transistor. Generally, it is found that the emitter-base voltage of germanium transistors is a few hundred millivolts, whereas the voltage is in the vicinity of 0.8 volt for silicon transistors. There are other practical aspects of biasing, particularly those relating to a-c considerations. These will be discussed in subsequent chapters.

Using Another Transistor for Emitter Stabilization. Inasmuch as the primary mechanism of bias stability is the enforcement of constant current in the emitter-base diode, other stabilization methods naturally suggest themselves. For example, instead of inserting resistance R_1 in the emitter lead, we can insert the collector circuit of another transistor which derives its stabilization from a conventional bias network (see Fig. 5-10). The collector circuit of

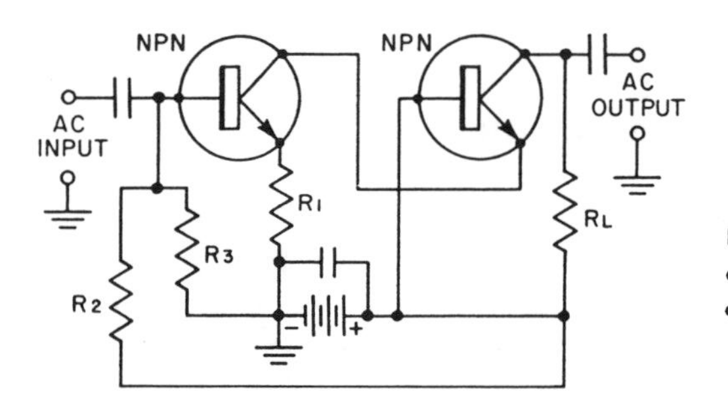

Fig. 5-10. Bias stabilization of transistor from constant-current source provided by preyious stage.

a stabilized transistor acts as a constant-current source, thereby qualifying as an appropriate substitution for R_1. Such a tandem arrangement of cascaded stages is particularly advantageous when high power levels are involved in the output stage.

Other Bias Network Modifications. In some instances, required circuit performance dictates modifications in biasing. A good example is the push-pull class-B stage. In class-B operation, the emitter-base biasing current is provided by the signal. When no signal is present, the transistors are "idling," and only I_{CO} flows in their collector collector circuits. When the half cycle of the signal is of appropriate polarity to provide forward bias to one of the transistors, that transistor draws collector current very nearly proportional to the instantaneous value of the signal current. The alternate transistor continues to idle at a collector current of I_{CO}, until the signal reverses in polarity, whereupon the conductive states of the two transistors exchange roles (see Fig. 5-11).

We note that, whereas the cutoff plate current of a tube is substantially zero, the transistor draws a residual collector current I_{CO}, regardless of cutoff bias applied to its base. Sometimes, a slight forward bias is applied to the transistors by a resistance connected from collector to base. This produces an idling current somewhat

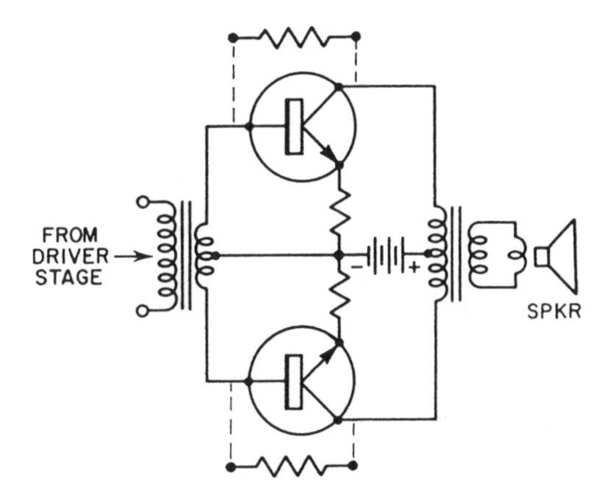

Fig. 5-11. A biasing arrangement for the push-pull class-B amplifier.

greater than I_{CO}. The objective here is to keep operation within relatively linear regions of transistor characteristics, thereby minimizing distortion. The resistances form a voltage divider in conjunction with the resistance of the secondary of the driver transformer. The slight amount of feedback produced does not generally alter circuit performance; if anything, decreased distortion at the crossover points results.

Bias in the Detector Circuit. Another case where it is necessary to dispense with conventional biasing networks is in the detector circuit of Fig. 5-12. The operation of this circuit, insofar as biasing is concerned, is similar to that of one transistor of the class-B push-pull pair. Because the collector is driven into conduction for only one polarity of the incoming signal, rectification or demodulation is accomplished.

Heat Dissipation. The matter of heat dissipation is logically included in our discussion of biasing. We bias the transistor in order to establish desired operating points. However, it is primarily under the stimulation of heat that drift in operating point occurs. The removal of heat is of paramount importance in the

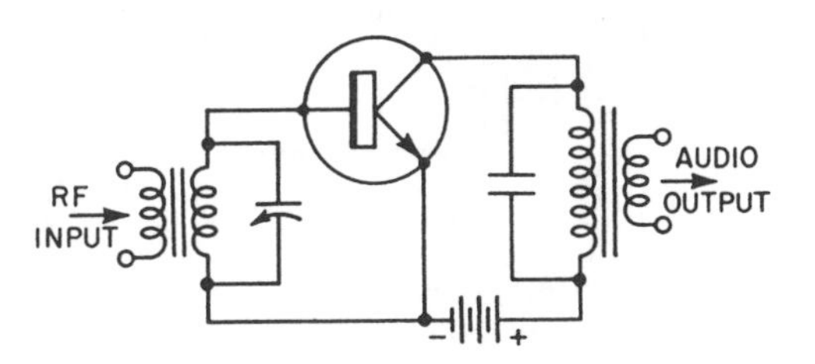

Fig. 5-12. Detector circuit.

operation of power transistors—generally those transistors which are called upon to deliver powers of several watts or more. Toward this objective, power transistors are constructed with their collector electrodes in physical as well as electrical contact with massive metallic plates or mounting studs. This construction enables efficient heat removal by the chassis or mounting plate, which is then known as a *heat sink*.

An approach to a true heat sink would require a large area of material having high thermal conductivity. In practice, the amount of metal used for draining heat from the transistor depends upon such factors as the allowable collector junction temperature, the efficiency of the thermal contact between transistor and sink, and the conditions of the air to which the sink is exposed. In order to circumvent circuit difficulties, it is often necessary to electrically insulate the transistor from the sink by means of a thin mica spacer. A certain amount of degradation in heat transfer is the unfortunate consequence of this technique.

Temperature- or current-sensitive resistance elements may be incorporated in the bias network in such a way that the collector current is prevented from increasing sufficiently to degrade circuit operation or from destroying the transistor. One such arrangement is shown in Fig. 5-13. Here, we note that R_3 is a thermistor. The resistance of the thermistor decreases with rise in temperature. When the circuit is operating in an environment having an elevated temperature, base-emitter forward bias is reduced, thereby counteracting the tendency toward higher collector current. The 100-ohm resistance R is inserted to make the effect of the thermistor somewhat less pronounced, inasmuch as its resistance decreases very

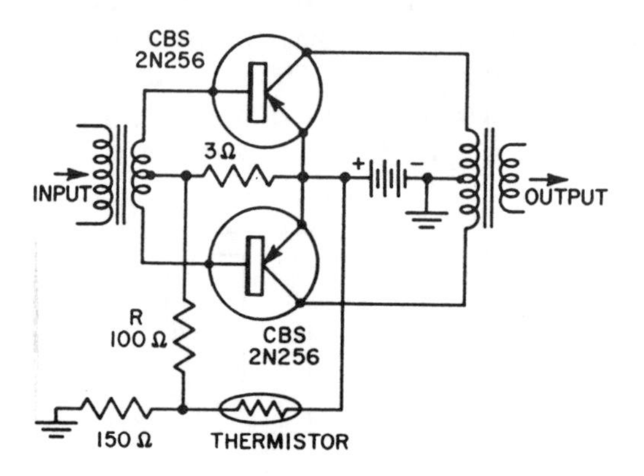

Fig. 5-13. Insertion of thermistor in power amplifier bias network to prevent degradation of performance and thermal runaway.

rapidly with temperature. R also isolates the input circuit from a-c distortion which might otherwise be introduced by the nonlinearity of the thermistor.

In d-c amplifiers, particularly those employing two or more cascaded stages, a high order of d-c stabilization is generally necessary. Variations in I_{CO}, in the input, or in an early stage, appear as signal variations to subsequent stages and receive the same amplification as would an intentional signal. This is essentially a bias problem and is often solved by a special circuit utilizing a pair of transistors. The transistors are connected in such a way that like disturbances in each produce no output signal. Conversely, only a difference in conductive state between the two yields an output signal. From this action the circuit derives its name as a *differential amplifier.* A vacuum-tube differential amplifier is shown in Fig. 5-14. It consists essentially of a cathode follower directly coupled to a grounded-grid amplifier.

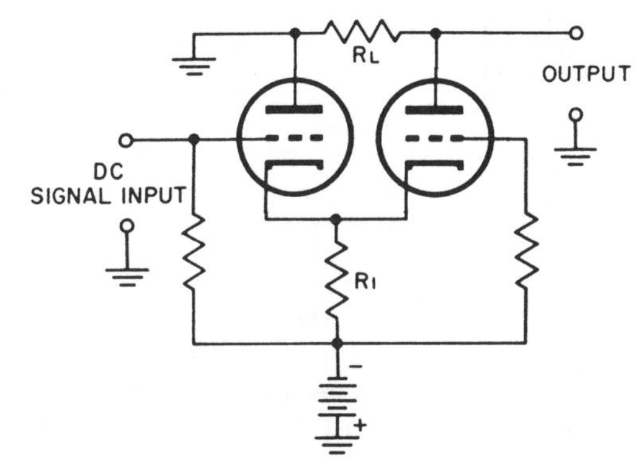

Fig. 5-14. The vacuum-tube differential amplifier.

Suppose for some reason, say an increase in filament current, both tubes tend to conduct more current. The increased conduction current of V_1 flowing through mutual cathode resistance R_1 increases the potential drop developed across R_1. This in turn increases the positive bias applied to the cathode of V_2, which *reduces* its plate current.

The converse sequence of events occurs in the event of a tendency for both tubes to assume a lower conductive state. We see that circuit action is such as to *counteract* the effect of *like* changes in the two tubes. In Fig. 5-15 the transistor version of this circuit is shown. It is assumed that the transistors are reasonably alike in their characteristics and are mounted close together so that both will undergo the same temperature variations.

In the event of an increase in ambient temperature, the increased I_{CO} in the two transistors will tend to cancel insofar as change in output voltage is concerned, the circuit mechanism being

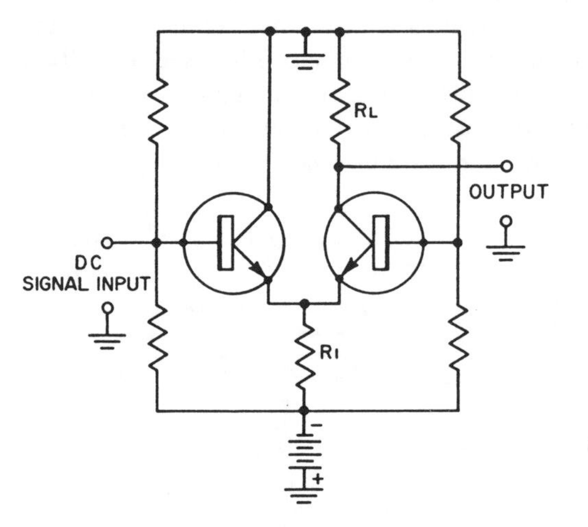

Fig. 5-15. The transistor differential amplifier.

essentially the same as in the vacuum-tube differential amplifier. We see that the useful property of the differential amplifier derives from the fact that output is obtained in response only to *differences* in conductive states of the two amplifying elements. The input signal is applied in such a way as to upset conduction balance, but the effects of changes in the I_{CO}'s of similar transistors will tend to cancel, as far as any change in output voltage is concerned.

6. the three transistor amplifier circuits

As with vacuum tubes, there are three different ways in which a transistor amplifier may be connected. One of these, the common-emitter circuit, is basically the same configuration as the most usual vacuum-tube amplifier in which the cathode is the common electrode to both input and output signals. A second transistor amplifier, the common-base configuration, is analogous to the grounded-grid amplifier in which the grid forms a mutual junction for both input and output signals. Finally the common-collector transistor amplifier is the transistor counterpart of the cathode-follower circuit. These analogies are shown in Fig. 6-1. In Fig. 6-2 the common-collector circuit is redrawn to more clearly show that the collector forms a mutual terminal for both input and output circuits.

THE COMMON-BASE CIRCUIT

In investigating the circuit properties of the three basic transistor amplifiers, it is expedient to start with the common-base circuit. The physical construction and the physics of its operation make this circuit a convenient reference for the characteristics of the other two configurations.

The first property we shall concern ourselves with in the common-base circuit is the current-gain factor α (alpha). α expresses

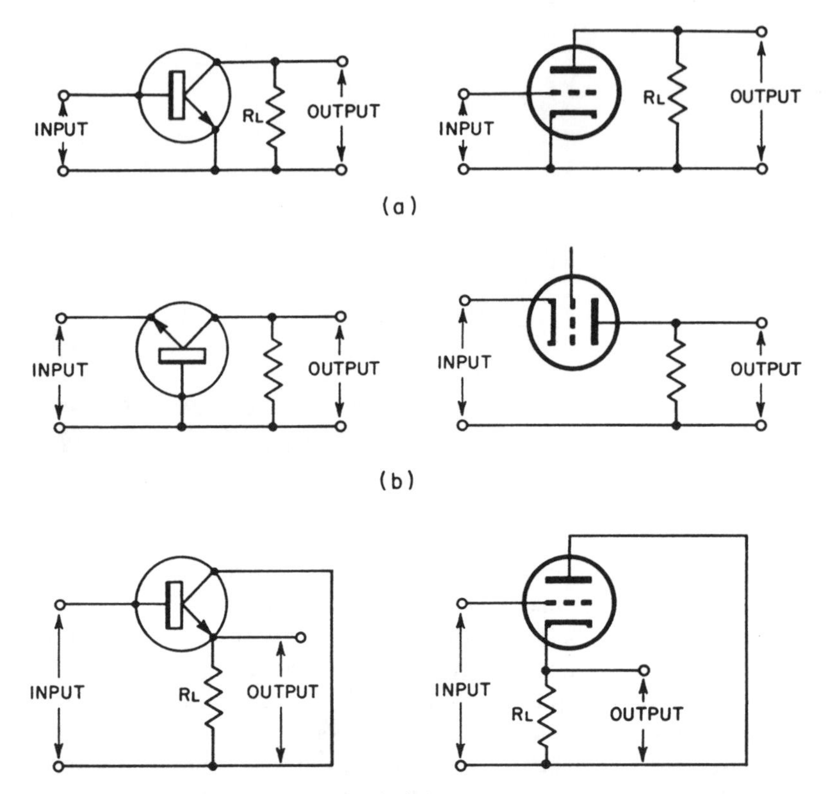

Fig. 6-1. The three transistor configurations and their vacuum-tube counterparts (a-c essentials *only*); (A) common emitter; (B) common base; (C) common collector.

the ratio of collector current to emitter current under short-circuit output conditions, that is, when $R_L=0$. Figure 6-3 shows the circuit arrangement for obtaining a. In junction transistors operating within factory-specified temperature range, a is a number close to but always less than 1. We must not infer from this that the

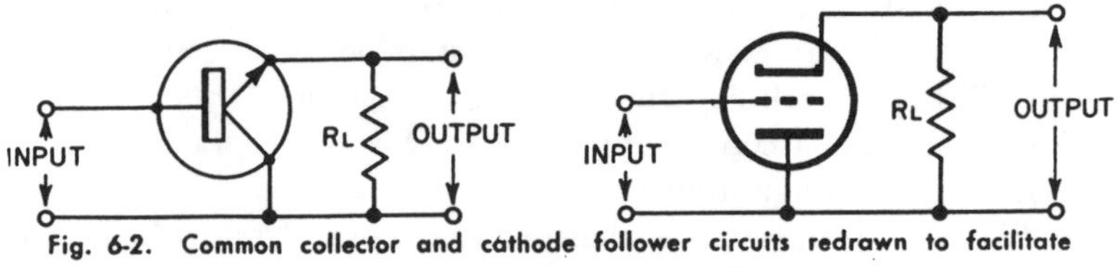

Fig. 6-2. Common collector and cathode follower circuits redrawn to facilitate comparison with other configurations.

common-base circuit does not provide power gain. Considerable power gain attends the operation of this circuit by virtue of its voltage amplification. Consider the equation

$$\alpha = \frac{\Delta I_c}{\Delta I_e}$$

where ΔI_c = change in collector-current resulting from closure of switch S_1

ΔI_e = change in emitter current resulting from closure of switch S_1

Alpha, The Common-Base Current-Amplification Factor. The current gain factor α of the common-base amplifier is a measurement of the physical fact that not all emitter carriers arrive at the collector. At low frequencies, say up to 1 kc, this loss, insofar as the external circuit is concerned, manifests itself as current in the base lead. Thus, it is very nearly true that the emitter current is equal to the sum of the base current and the collector current.

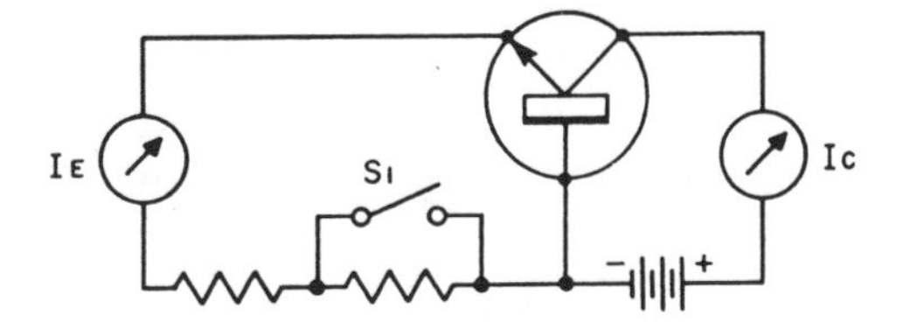

Fig. 6-3. Circuit for measuring α in common-base configuration.

The two important causes of base current are (1) emission of minority carriers from base to emitter and (2) recombination of majority carriers in the base region. The latter phenomenon is analogous to interception of electrons by a positive grid in the vacuum tube.

At high frequencies another phenomenon exists. This is a *transit-time effect.* The mobility of the carriers and the thickness of the base region impose a limit on the speed at which an injected carrier can diffuse to the collector region. We see that, when the frequency is sufficiently high, arrival of carriers in the collector region may no longer be in time phase with the input signal. This in itself would not be of great consequence if a "package" of charge carriers all crossed the base region together. There would then exist a time delay between output and input signals which would not limit high-frequency operation at as low frequencies as actually occurs in transistors. Inasmuch as the diffusion process of carrier transit through the base is characterized by "randomness," the arrival of injected carriers in the collector region has something of a smeared rather than a "stacatto' effect. As long as the bulk of the relatively late arriving carriers make their transit of the base region in a time which is small compared to the duration of a cycle of the operating frequency, α will remain substantially at its low-frequency value.

As the operating frequency is raised, this is no longer the case, for then it is no longer possible for the majority of injected carriers to cross the base region in sufficient time to contribute to a collector replica of the emitter signal. Consequently, we see that α must decrease at high frequencies. By definition, the frequency at which alpha is 71% of its low-frequency value is the α cutoff frequency. The cutoff is not a sharp one as in certain filters, but the output of a common-base amplifier is approximately limited to this frequency insofar as flat frequency response is concerned. Narrow-band tuned amplifiers and oscillators can operate at reduced output considerably beyond α cutoff frequency.

Common-Base Circuit Properties. The salient characteristics of the common-base amplifier are as follows: This configuration does not reverse the phase of the signal undergoing power amplification. It does not produce current amplification, but provides appreciable voltage and power gain. Its input impedance is the lower extreme encountered in the three circuits, whereas its output impedance is the highest. Consequently, the cascading of common-base stages by resistance–capacitance coupling results in such a severe impedance mismatch that usable power gain is not obtained. Common-base stages can however be cascaded by appropriate impedance-matching transformers. The common-base stage is capable of providing a power gain in excess of that obtainable from the same transistor connected as a common collector stage, but not as great as can be obtained with the same transistor connected in the common emitter circuit. Of the three configurations, the common-base stage provides the best isolation insofar as concerns the effect of the output circuit on the input circuit.

Collector Current in Common-Base Circuit. An interesting property of the common-base circuit derives from the fact that collector current is virtually at its operating value even when the collector voltage is zero. This enables a very close approach to

50% efficiency for class-A amplifier operation. The collector characteristics illustrated in Fig. 6-4 makes this clear. Further exposition of common-base characteristics will continue in the ensuing discussion of the *common-emitter* amplifier. A better insight into the properties of both circuits is made possible by *comparison* of the two, for it will be seen that a unique relationship exists between them.

THE COMMON-EMITTER CIRCUIT

The common-emitter amplifier, which resembles the most frequently encountered tube circuit, is generally easier to work with than the common-base configuration. The input and output impedances both have more moderate values. As a result, despite

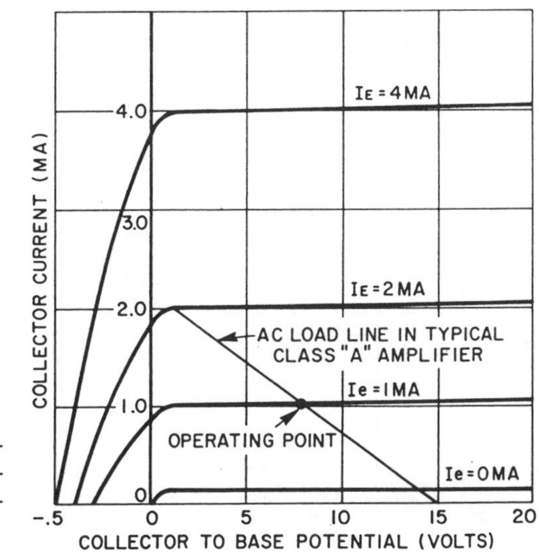

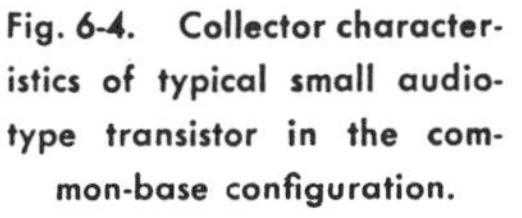

Fig. 6-4. Collector characteristics of typical small audio-type transistor in the common-base configuration.

the impedance mismatch which does exist, usable gain is accumulated from cascaded common-emitter stages which are resistance-capacitance coupled. The power gain of the common-emitter stage is the highest of the three amplifier circuits. This is because the common-emitter circuit develops *both* current and voltage gain, whereas the other two configurations provide voltage or current gain only.

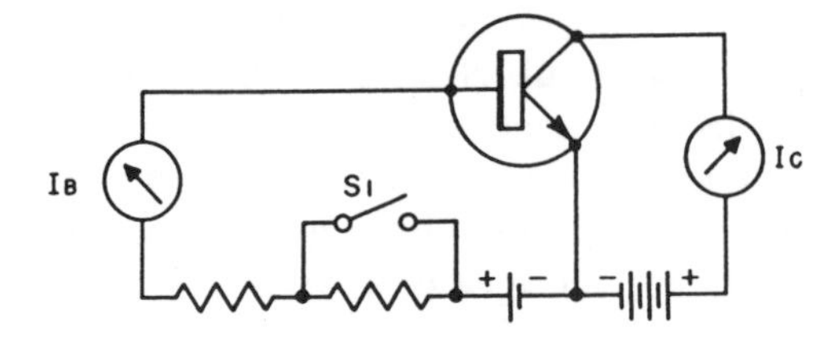

Fig. 6-5. Circuit for direct measurement of current-gain factor β in common-emitter amplifier.

The important properties of the common-emitter circuit are derived from the counterpart characteristics of the common-base circuit. For example, β, the current gain factor of the common-emitter circuit, is obtained from α; thus $\beta = \alpha/(1 - \alpha)$. Transposition of this relationship yields the useful formula

$$\alpha = \frac{\beta}{\beta + 1}$$

We see that simple calculations permit derivation of either α or β if the value of one of these parameters is available. β is defined as the collector circuit magnification of a small change in base

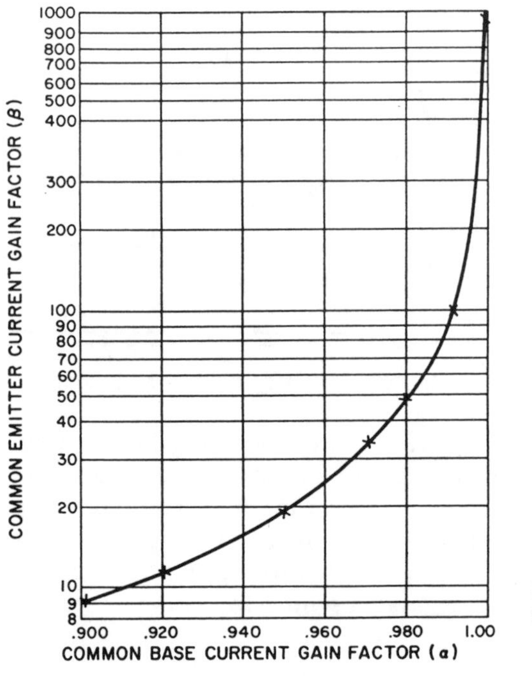

Fig. 6-6. Relationship between current gain factors α and β.

current. The circuit for directly measuring β is shown in Fig. 6-5. The relation between α and β is shown in Fig. 6-6.

$$\beta = \frac{\Delta I_c}{\Delta I_b}$$

where ΔI_b = change in base current due to closure of S_1
ΔI_c = corresponding change in collector current

Although α is a small number, less than 1, β is a large number which increases rapidly as α approaches 1. The higher current gain factor of the common-emitter circuit could have been anticipated from the circuit conditions of the common-base stage.

In Fig. 6-7 the significant fact is that the two currents in the base lead flow in *opposite* directions. The net base current is then the difference between emitter and collector current. Inasmuch as emitter and collector current are nearly the same, the base current must consequently be much *smaller* than the emitter current, which is to say we have more collector current magnification with respect to base current than with respect to emitter current.

$I_b = I_c - I_e$ assuming I_{CO} is negligible compared to I_c

$I_e >> I_b$

We note, of course, that one of the prime differences between common-base and common-emitter amplifiers is the signal injection point.

RELATING COMMON-BASE AND COMMON-EMITTER CIRCUITS

Conversion Factor. We have seen that β is a value which is $1/(1-\alpha)$ times as great as α. This multiplying factor transforms other common-base parameters to those pertaining to the common-emitter circuit. For example, the β cutoff frequency is a frequency equal to the α cutoff frequency divided by this factor; that is,

$$\beta \text{ cutoff frequency} = \frac{\alpha \text{ cutoff frequency}}{1/(1-\alpha)}$$

or more simply, α cutoff frequency times the factor $(1-\alpha)$.

We see that the β cutoff frequency is much less than the α cutoff frequency. β cutoff frequency is defined as the frequency at which short-circuit collector current is 71% of its value at low frequencies, say several hundred cycles. Although this implies poorer frequency response, it is often possible to extend frequency response by negative feedback. This is generally practical because, as explained in the discussion on feedback, the improved impedance match often obtained by appropriate feedback connections can regain the power loss which might otherwise be the price of

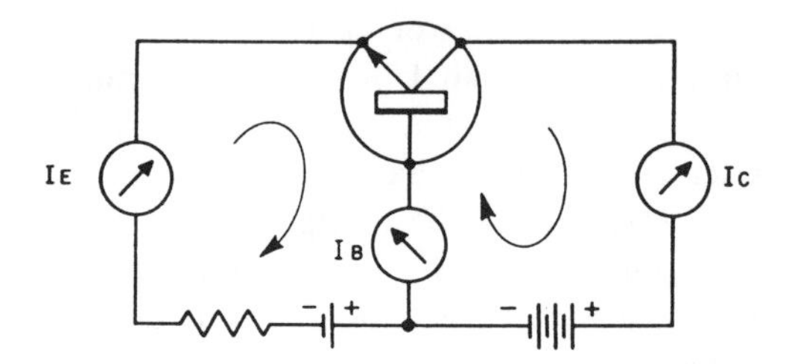

Fig. 6-7. Insight into common-emitter operation is provided by consideration of base current in the common-base amplifier.

feedback. Even where this may not apply, the nonfeedback power gain of the common-emitter circuit is high, and a power loss can often be spared.

In still other cases, the β cutoff frequency is well above the maximum operating frequency; so the full power gain inherent in this amplifier may be obtained together with satisfactory frequency response.

Input Impedance. Another modification by the factor $1/(1-\alpha)$ is encountered in the input impedance to the two amplifier circuits. The input impedance to the common-emitter circuit is very nearly equal to $1/(1-\alpha)$ times the common-base input impedance. The output impedance of the two configurations is also approximately related by the factor $1/(1-\alpha)$ with the common-emitter output impedance being the lower in value. Thus, for most practical purposes we may use the relationship that common emitter output impedance is approximately given by

$$\frac{\text{common-base output impedance}}{1/(1-\alpha)}$$

or more simply, common-base output impedance times the factor $(1-\alpha)$.

Voltage Gain. For most circumstances encountered in practice, the voltage gain of both circuits may be considered to be numerically the same but with one difference: that is, the common-emitter circuit produces phase reversal of the signal undergoing amplification, whereas the common-base amplifier does not. For both circuits the numerical voltage gain is approximately

$$\frac{r_L(a)}{r_e + r_b(1 - a)}$$

The values of r_e and r_b are dependent upon emitter-base current and may be approximated from the manufacturer's specifications.

For many applications involving small general-purpose or audio-type transistors, we may gain some notion of a "ball-park" value of voltage amplification by estimating r_e as follows:

$$r_e \cong \frac{15}{I_e}$$ for germanium alloy junction structures

$$r_e = \frac{30}{I_e}$$ for grown junction structures (both germanium and silicon)

In both instances I_e is the emitter current in milliamperes. The base resistance, r_b, may be assumed to be within the 250- to 500-ohm range for the purposes of such an estimate.

Power Gain. Finally, the power gain of the common-emitter amplifier is $1/(1 - a)$ times that of the common-base amplifier. The approximate formulas for the two circuits are as follows:

$$\text{Common-base power gain} = \frac{a^2 r_L}{r_e + r_b(1 - a)}$$

$$\text{Common-emitter power gain} = \frac{1}{1 - a} \times \frac{a^2 r_L}{r_e + r_b(1 - a)}$$

Another formula for determination of power gain involves terms which are somewhat easier to measure or estimate.

$$\text{Power gain} = \left(\frac{I_{out}}{I_{in}}\right)^2 \left(\frac{r_L}{r_{in}}\right)$$

This formula is applicable to all three amplifier circuits. I_{in} is the current measured in the base lead in the common-emitter (and common-collector) circuit. For the common-base circuit, I_{in} is the current measured in the emitter lead. If r_{in} is the input impedance of the transistor, the power gain is then that corresponding to the transistor in conjunction with the particular load, r_L. If r_{in} is the net input impedance due to transistor and bias network, the power gain is then that corresponding to the actual amplifier circuit.

THE COMMON COLLECTOR CIRCUIT

The common-collector configuration is the third transistor-amplifier circuit. This configuration is analogous to the vacuum-tube cathode follower, which in essence is a *common-plate amplifier.*

It has been shown in the common-base and common-emitter circuits that, for a given collector current, a nearly equal current exists in the emitter lead, but a much smaller current flows in the base lead. We may say that there is current amplification in the emitter circuit with respect to the base circuit. It is not to be inferred that such amplification exists in a simple p-n junction diode or in the common emitter circuit with no collector current. Although in the common-collector configuration the output is derived across a resistance inserted in the emitter lead, the current amplification is the consequence of transistor action and is due to flow of current in the emitter-collector circuit.

As in the common emitter circuit, the multiplying factor $1/(1-a)$ plays a prominent role in defining the operating parameters. The important relationships in the common-collector circuit are as follows:

$$\text{Current gain} = \frac{1}{1-a} \quad \text{which, we note is approximately } \beta$$

$$\text{Input impedance} = \frac{1}{1-a} r_L$$

We see that a high input impedance requires a high impedance load. It should be borne in mind that the load being driven exerts a shunting effect on the actual resistance inserted in the emitter lead.

$$\text{Output impedance} = r_e + \frac{r_g + r_b}{1/(1-a)}$$

or

$$\text{Output impedance} = r_e + (r_g + r_b)(1-a)$$

where r_g is the impedance of the signal source. We note that, as in the other amplifiers, r_L does not appear in our simplified formula for output impedance. The output impedance is the impedance seen by the load looking back into the transistor.

$$\text{Voltage amplification} \cong 1 \cong \frac{r_L}{r_e + r_b(1 - a) + r_L}$$

The common-collector circuit is, in a sense, the opposite of the common-base circuit. The common-collector circuit provides current amplification but not voltage amplification, whereas the converse situation prevails with the common-base circuit. This is reflected in the impedance of the two circuits which are of opposite extremes. On the other hand, a mutual property of both configurations is that no phase reversal is imparted the signal undergoing power amplification.

The cutoff frequency of the common-collector amplifier is substantially the same as that of the common-emitter amplifier and is obtained from the relationship

$$\text{Common collector cutoff frequency} = \frac{a_{CO}}{1/(1 - a)} \quad \text{or} \quad a_{CO}(1 - a)$$

where a_{CO} is the cutoff frequency of the common-base amplifier. For most practical purposes, a useful approximation is

$$\text{Common-collector cutoff frequency} = \frac{a_{CO}}{\beta}$$

COMPARISON OF CIRCUIT PARAMETERS IN THE THREE CONFIGURATIONS

Table 6-1 presents the relative values associated with the three amplifier circuits. Further insight into the nature of the transistor amplifier may be gained by inspection of the typical parameter values shown in Table 6-2. The value ranges indicated are not necessarily related, but have been chosen from consideration of many circuits to convey a practical "feel" for the transistor amplifier. As with vacuum tubes, transistors are made with different power-handling capabilities. High-power transistors will not have the value ranges indicated in Table 6-2. Most significantly, their input and output impedances will generally be below 100 ohms, and their a cutoff frequencies generally do not exceed 100 kc.

However, the correlation of relative values with configuration shown in Table 6-1 is still valid.

TABLE 6-1.
RELATIVE VALUES OF TRANSISTOR CHARACTERISTICS FOR THE THREE AMPLIFIER CONFIGURATIONS

Characteristic	*Common base*	*Common emitter*	*Common collector*
Analogous vacuum-tube circuit	Grounded grid amplifier	Grounded-cathode amplifier	Cathode follower
Maximum power gain	Moderate	Highest	Lowest
Current gain	Near unity (α)	High (β)	Highest ($\beta + 1$)
Voltage gain	High	High	Near unity
Phase of output signal relative to input signal	No phase reversal	Phase reversal	No phase reversal
Half-power frequency (α or β cutoff frequency)	Highest	Low	Low
Input impedance	Lowest	Moderate	Highest
Output impedance	Highest	Moderate	Lowest

Cascaded Stages Using Resistance-Capacitance Coupling. Although transformer coupling enables cascading of any combination of transistor amplifier configurations, we generally do not have this design liberty with resistance-capacitance coupling. For example, it would not be profitable to make a resistance-capacitance-coupled amplifier using two common-base stages. The input impedance of the second stage would be much too low to permit the first stage to develop reasonable power gain. The same would be true in a common-emitter–common-base circuit using resistance-capacitance coupling. In most cases of resistance-capacitance coupling, optimum power gain will not be obtained due to mismatch. However, some combinations are much better than others. A usable system of cascaded stages results from employing resistance-capacitance coupling between common-emitter stages. Despite considerable power loss due to impedance mismatch, the inherent high power gain of the common emitter circuit permits such a cascaded arrangement to develop practically any required power amplification by the employment of an appropriate number of stages.

The Common Collector Stage as an Impedance Matcher. Where space, weight, or economic factors exclude the use of transformer coupling, the common-base output amplifier can often be efficiently driven by a common-collector stage. In turn, the common collector stage can be preceded by either a common-base or common-emitter input stage. In such a three-stage amplifier, the common-collector stage performs the role of an impedance-matching device. This it is able to do by virtue of its relatively high input impedance and low output impedance. It should be appreciated, that, whereas in a vacuum-tube arrangement, a cathode follower might be considered a "wasted" stage because it provides no voltage gain, the common-collector stage provides a usable current gain in the transistor amplifier. The common-collector stage is capable of providing more drive than a transformer to a power output stage.

Circuit Considerations for High Input Impedance in Common-Collector Stage. A fact often overlooked in both the vacuum-tube cathode follower and the transistor common-collector stage is that high input impedance does not naturally ensue regardless of the load seen by the output. If the load is a low impedance, the input impedance will not be high. In the cathode follower this relationship is perhaps of less consequence because the input impedance to the grid circuit of a tube operating as a class-A amplifier is inherently high. However, with the transistor, the input diode, operating in its forward-conduction region, has an inherently low

**TABLE 6-2.
REPRESENTATIVE VALUES LIKELY TO BE ASSOCIATED WITH A TYPICAL SMALL AUDIO-TYPE JUNCTION TRANSISTOR**

Characteristic	*Common base*	*Common emitter*	*Common collector*
Power gain, P_{out}/P_{in}	600–1200	8000–12,000	12–50
Current gain, I_{out}/I_{in}	0.920–0.980	12–50	12–50
Half-power frequency range, kc	800–1200	40–300	40–300
Voltage gain, E_{out}/E_{in}	200–600	15–50	15–50
Input impedance, ohms	40–80	600–1200	5–50 K*
Output impedance, ohms	250–750 K	40–80 K	100–400

*Very dependent on load resistance.

impedance. This impedance can be effectively raised by means of appropriate feedback applied to the common-emitter circuit. This will be discussed in subsequent paragraphs.

Signal Dissipation by Bias Network. A direct approach to the a-c amplifier is the circuit depicted in Fig. 6-8. There is the possibility that the divider network R_2 and R_3 can dissipate an undesirable portion of the incoming power. This situation will be encountered when the impedance of the driving source is high, say two or more times the impedance seen looking into the input

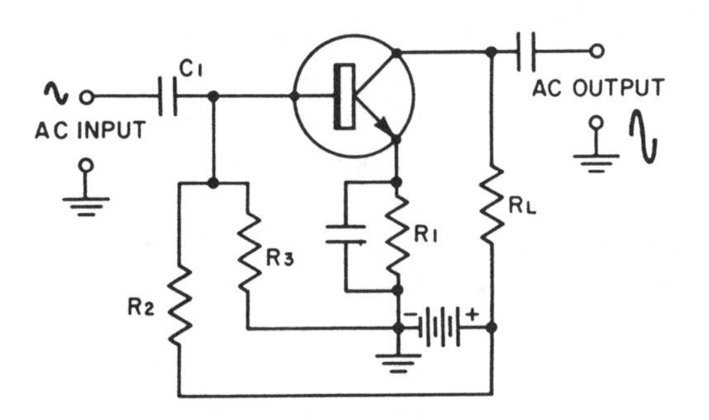

Fig. 6-8. A-c amplifier using capacitance coupling.

terminals of our amplifier. Under such conditions, it is often possible to modify the bias network so that this shunting effect will be reduced; that is, both R_2 and R_3 can be raised by a factor of several or more times. In most practical situations stability will not be greatly degraded by this design modification inasmuch as R_1, the emitter resistance, generally provides sufficient stabilization.

Another facet of avoiding excessive signal shunting effect has to do with R_L. It is not profitable to make resistance R_L high unless the loading effect of the subsequent stage is accounted for in estimating the a-c gain of our amplifier. It follows that transformer coupling is a means whereby the impedance of one stage can be matched to the impedance of another. However, it may still be necessary to consider the shunting effect of the bias network.

Overcoming Signal Dissipation by Bias Network. Figure 6-9 shows one way in which transformer coupling can be incorporated into a transistor amplifier. It must be appreciated that the objective, insofar as efficient transformation of power is concerned is to match the output impedance of one stage to the input impedance of the following stage. When the input impedance is substantially comprised of bias resistances, a power loss is inevitable.

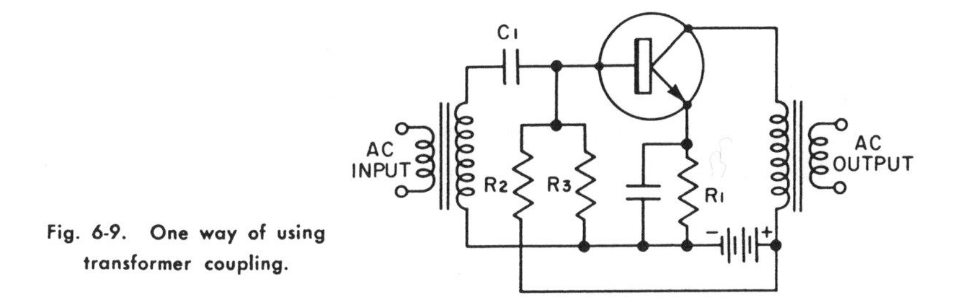

Fig. 6-9. One way of using transformer coupling.

This cannot be avoided by matching into the net effective input impedance. Such a loss may be reduced by increasing the values of both R_2 and R_3.

An alternative scheme of connecting the transformer avoids this loss altogether. In Fig. 6-10 the d-c bias to the transistor input diode is series-fed rather than shunt-fed, as in the circuit of Fig. 6-9. Both circuits require a blocking capacitor C_1. In the circuit of Fig. 6-10 no a-c shunting effect is provided by bias resistances R_1 and R_2. Efficient power transformation prevails inasmuch as the transformer feeds only the input diode of the driven stage.

COMPLEMENTARY SYMMETRY

The existence of transistors of two conduction types, n-p-n and p-n-p, make possible interesting and useful circuit combinations which have no counterpart in tube practice. Figure 6-11 shows a direct-coupled amplifier using an n-p-n transistor as driver for a p-n-p power output stage. Both stages employ the common-emitter configuration. A little consideration reveals that a conflict in bias-

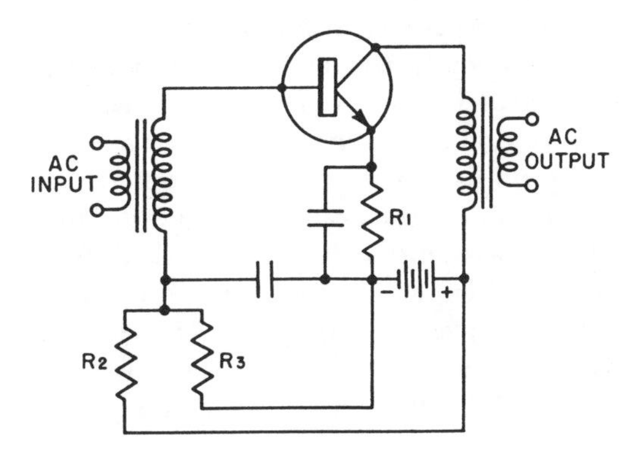

Fig. 6-10. Alternative transformer connections.

ing polarity would prevent such direct coupling of common-emitter stages using transistors of like conductivity type. When the transistors are opposite types, the same current biases the output diode of the driver stage and the input diode of the power stage. The arrangement depicted in Fig. 6-11 is a fortunate one

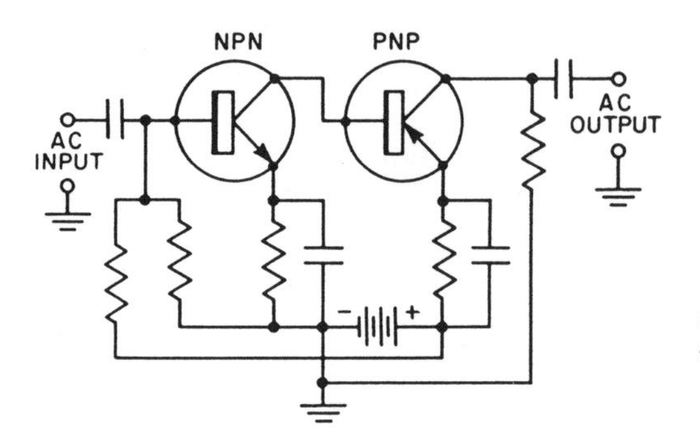

Fig. 6-11. Transistor amplifier employing principle of complementary symmetry.

inasmuch as there is a wider selection of p-n-p power transistors available than of n-p-n power units. However, the principle of the circuit remains valid for the case of a p-n-p first stage and n-p-n second stage.

Push-Pull Class-B Amplification by Complementary Symmetry. The principle of complementary symmetry may be used in a simple circuit which functions as a push-pull class-B output amplifier, yet dispenses with the need for phase inverter as well as a push-pull output transformer. This circuit is formed by simply connecting otherwise similar n-p-n and p-n-p transistors essentially in parallel. Such an arrangement is shown in Fig. 6-12. With no input signal neither transistor conducts, because no forward bias is provided. On the positive excursion of the a-c input signal the n-p-n transistor becomes conductive and delivers power to the load. On the negative excursion of the a-c input signal the p-n-p transistor assumes this role. Thus over a complete cycle, the circuit action is essentially similar to that of conventional push-pull class-B amplifiers.

Input Impedance Dependence on Bias Current. The impedance of the base-emitter diode decreases as forward current is increased. Thus, when we cite an input impedance for an a-c signal, we imply an average impedance. If the amplitude of the applied a-c is not too great, the average input impedance will be substantially governed by the bias current and voltage employed to establish d-c forward conduction. The instantaneous input impedance decreases with input driving current. It follows that the applied a-c wave must suffer distortion. This distortion is generally most severe in the common-base circuit. It can be diminished by inserting a resistance in series with the emitter lead, or by driving from a source having a higher impedance than that corresponding to a match.

Such techniques introduce losses, but can often be profitably used to decrease distortion. In the common-emitter circuit, an unbypassed resistance is frequently inserted in the emitter lead

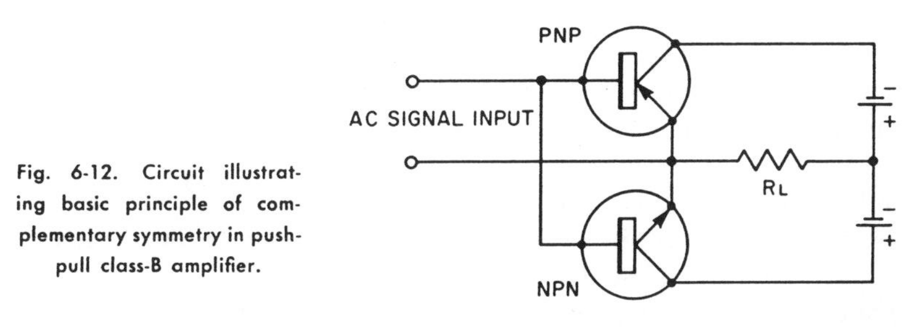

Fig. 6-12. Circuit illustrating basic principle of complementary symmetry in push-pull class-B amplifier.

for the purpose of providing negative feedback. Such feedback raises the effective input impedance by a factor greatly in excess of that which would result from simple ohmic addition of the inserted resistance.

This situation is even better in the common-collector circuit where, ideally, 100% inverse feedback can raise the effective input impedance to a very high value. Of course, we start with a better situation in the common-emitter and common-collector circuits than in the common-base circuit. This is because a smaller input current suffices to produce a given collector current. Except for exacting requirements, the nonlinearity of the input circuit is generally of practical consequence only in the common-base amplifier.

NEGATIVE FEEDBACK IN TRANSISTOR AMPLIFIERS

Both the theoretical and practical aspects of feedback can constitute a voluminous study. We shall restrict our investigation of feedback to generalities which are, however, particularly relevant to transistor amplifiers. From vacuum-tube circuitry, we recall that negative feedback is extensively employed for the purpose of reducing distortion. Feedback accomplishes the same result in transistor amplifiers, but should not be used without regard to the input and output impedance changes inevitably produced. One reason why this consideration is generally overlooked in vacuum-tube amplifiers is that the input impedance of a tube is, insofar as the majority of practical circuits is concerned, infinite. Thus it is of little practical consequence that feedback can increase or decrease this very high impedance by, say a factor of 10.

In a transistor, however, the input impedance is relatively low, often between several dozen and several kilohms. This necessitates consideration of impedance matching in order to avoid excessive loss in power gain. Inasmuch as two or more amplifying stages are often cascaded, we see that the relative input and output impedances of individual stages are of great importance in *R-C*-coupled amplifiers.

It is usually profitable to manipulate these impedances by feedback in the direction of at least an approximate match. In so doing, power gain will be used up by the feedback but will be virtually recovered by the improved match. The net result is about the same power gain as without feedback, but with the desirable features provided by feedback. These are reduced distortion, increased a-c stability, diminished importance of transistor characteristics, and improved frequency response. When transformer coupling is used, the same performance improvements are attained, but one must provide for an attendent reduction in the maximum possible power gain.

Effect of Feedback on Current Gain. Negative feedback in a transistor amplifier is described by the modifying factor $(1 + K)/(1 + KG)$ where G represents current gain without feedback and K is the percentage of a-c output current returned to the input through the feedback network. This modifying factor becomes more meaningful when we employ it to express the change in operating parameters brought about by the feedback. For example, the current gain of a feedback amplifier is given by

$$G \frac{1 + K}{1 + KG}$$

This relationship provides us with interesting insight into the stabilizing nature of negative feedback. In most practical feedback amplifiers, particularly those involving a feedback path encompassing two or more cascaded stages, K is small enough compared to 1 to merit dismissal from the numerator for the sake of simplicity.

Our simplified expression for current gain with feedback is therefore

$$G\frac{1}{1+KG}$$

Now, let us suppose that by virtue of high nonfeedback gain, the product KG is much larger than 1. This enables us to make an additional simplification with little deviation from accuracy. Our current gain with feedback now is very nearly $G(1/KG)$ or G/KG, which yields the ultimate simplification $1/K$. That is:

$$\text{Current gain with feedback} = \frac{1}{K}$$

Significance of Feedback Equation. The most significant aspect of this relationship centers about the parameters which are not contained therein. We note that the current gains of individual transistors are not involved. Nor are such factors as operating voltage or current. The natural conclusion is that the negative feedback has stabilized the gain of our amplifier. In practice, this is indeed the case; the higher the product KG, the more immune is the amplifier to transistor circuit and operating voltage variations.

Effect of Negative Feedback on Frequency Response. The factor $1/(1 + KG)$ (or $1/K$ when KG is large compared to 1) governs other parameters as well as current gain. For example, the frequency response with feedback is extended over the nonfeedback response as follows: Let A = the half-power point at the low-frequency end without feedback. Let B = the half-power point at the high-frequency end without feedback. Then the new low-frequency half-power point when feedback is applied is the frequency given by

$$A\frac{1}{1+KG} \quad \text{or} \quad \frac{A}{1+KG}$$

The new high-frequency half-power point when feedback is applied is the frequency given by

$$\frac{B}{1/(1+KG)} \quad \text{or} \quad B\,(1+KG)$$

The Different Ways of Providing Feedback Paths. We shall now concern ourselves with the various types of negative feedback and their relative effect on input and output impedances. To a first approximation these impedances are raised or lowered by the factor $1/(1 + KG)$. Accordingly, henceforth we shall simplify our discussion by merely stating whether input or output impedance is raised or lowered. The impedances are those looking into the input or output of the transistor divorced from bias resistance or load impedances.

We begin our feedback investigation by considering the "jack-of-all-trades" circuit shown in Fig. 6-13. This circuit con-

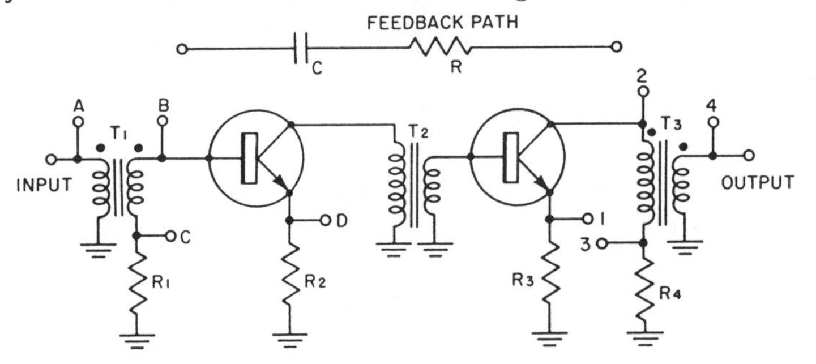

Fig. 6-13 A-c essentials of two-stage amplifier showing possible points for connection of feedback path. Refer to Table 6-3 for significance of various connections.

tains a number of designated points to which we can connect a feedback path consisting of a resistance and capacitance in series. We assume that the capacitor is large and functions only as a means of blocking the flow of d-c current. This circuit, in conjunction with Table 6-3, provides the impedance-changing information we need to determine the various ways in which the feedback path may be connected to provide return of a fraction of the a-c output to the input circuit. (Not all combinations are shown, inasmuch as we are restricting our discussion to negative feedback only.)

Feedback Derivation. We should know that there are two basic ways of deriving feedback from the output circuit and, likewise, two basic methods of applying or returning feedback to the input circuit. Let us first consider the derivation of feedback.

If we derive our feedback current in such a way that it is proportional to the a-c output voltage, we describe such feedback as being *voltage-derived.* Feedback derived from points 2 and 4 in Fig. 6-13 is voltage-derived.

TABLE 6-3.

THE EFFECT OF NEGATIVE FEEDBACK CONNECTIONS ON INPUT AND OUTPUT IMPEDANCE OF A TRANSISTOR AMPLIFIER

Point of Feedback Derivation in Fig. 6-13 (1)	*Point of Feedback Application in Fig. 6-13* (2)	*Circuit Conditions* (3)	*Description of Feedback* (4)	*Output Impedance* (5)	*Input Impedance* (6)
1	*A*	T_2 connected for *no* phase reversal	Current-derived; shunt-applied	Raised	Lowered
1	*B*				
1	*C*		Current-derived; series-applied	Raised	Raised
1	*D*	T_2 connected for phase reversal			
2	*A*	T_2 connected for phase reversal Also, R_3 and R_4 are shorted or bypassed	Voltage-derived; shunt-applied	Lowered	Lowered
2	*B*				
2	*C*		Voltage-derived; series-applied	Lowered	Raised
2	*D*	T_2 connected for no phase reversal			
3	*A*	T_2 connected for phase reversal	Current-derived; shunt-applied	Raised	Lowered
3	*B*				
3	*C*		Current-derived; series-applied	Raised	
3	*D*	T_2 connected for *no* phase reversal			
4	*A*	T_2 connected for phase reversal Also, R_3 and R_4 are shorted or bypassed	Voltage-derived; shunt-applied	Lowered	Lowered
4	*B*				
4	*C*		Voltage-derived; series-applied	Lowered	Raised
4	*D*	T_2 connected for no phase reversal			

If feedback current is obtained from the output in such a way that it is proportional to the a-c output current, such feedback is described as being *current-derived.* Feedback derived from points 1 and 3 in Fig. 6-13 is current-derived feedback. Given percentages of either type of feedback derivation are about equally beneficial in reducing distortion and in stabilizing the a-c gain of the amplifier. However, they have opposite impedance-changing effect on the output of the amplifier. Voltage-derived feedback lowers the output impedance. Current-derived feedback raises the output impedance.

Return of the Feedback Path. Now let us investigate the two ways of applying or returning feedback to the input of the amplifier. If the feedback current is returned in such a way that it acts in parallel with the input signal, such feedback is said to be *shunt-applied.* Feedback current returned to point *A* or *B* of Fig. 6-13 is shunt-applied.

If the feedback current is returned in such a way that it acts in series with the input signal, it is said to be *series-applied.* Feedback current returned to point *C* or *D* is of this variety.

Given feedback percentages confer approximately equal improvements in distortion reduction and a-c-gain stabilization, whether shunt- or series-applied. However, the two methods of feedback application produce opposite changes in the input impedance to the amplifier. Shunt-applied feedback lowers the input impedance. Series-applied feedback raises the input impedance.

An Example of Impedance Modification with Feedback. An interesting consequence of impedance manipulation by feedback is found in resistance-capacitance-coupled amplifiers. Consider the two-stage cascaded emitter amplifier shown in Fig. 6-14. To begin with, assume that R_2 is shorted out and the network R_1-C_1 is

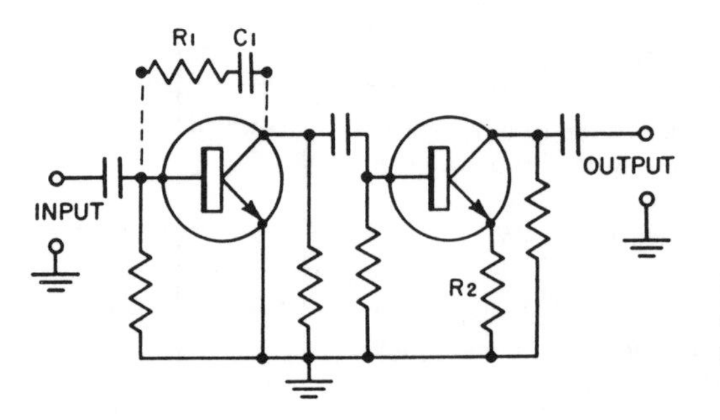

Fig. 6-14. A-c essentials of a two-stage amplifier which can be improved by load feedback.

not in the circuit. The output impedance of the first stage may be 10 to 100 times as great as the input impedance to the second stage. A considerable loss in power gain results from such a drastic mismatch.

If we connect the network R_1-C_1 from collector to base of the first stage, we lower the output impedance of this stage, thereby providing a better impedance match to the second stage. The over-all power gain of the two-stage amplifier will not thereby increase because negative feedback decreases power gain. However, the first stage will now have the stabilizing and distortion-reducing benefits of negative feedback.

If we are to obtain optimum results from this technique, we must provide for the lowered impedance in the input of the first

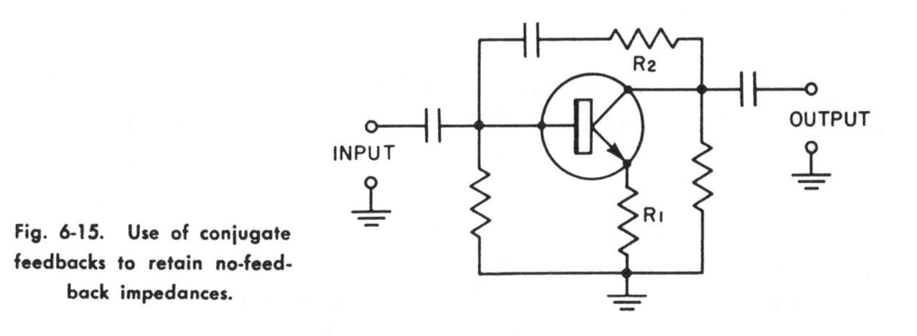

Fig. 6-15. Use of conjugate feedbacks to retain no-feedback impedances.

stage. A lower a-c driving impedance is now necessary to avoid gross impedance mismatch at the input.

Other Feedback Techniques for Impedance Control. An alternate procedure is to insert R_2 in the emitter lead of the second stage. This raises the input impedance of the second stage, which is again a change in the direction of impedance match between the two stages. In order to prevent unnecessary sacrifice of power gain, we must now work into a higher impedance output load, inasmuch as R_2 also raises the output impedance of the second stage.

Many other feedback techniques can be conjured from the information contained in Table 6-3. An example is the single common-emitter stage with conjugate feedback paths shown in Fig. 6-15. In this arrangement, R_1 raises both input and output impedance, whereas R_2 lowers both impedances. These two feedbacks may be proportioned so as to retain the "no-feedback" input and output impedances where this is desirable, say in carrier telephone applications where it is often necessary to insert an amplifier between designated impedances.

Feedback in the Common-Base Amplifiers. Our discussion of feedback has centered about the common-emitter stage. The distortion and instability of the common-base configuration are generally low enough so that feedback is not needed. However, there are applications where such feedback may prove beneficial. We cannot simply insert an unbypassed resistance in the base lead or connect a resistance and a-c blocking capacitor from collector to emitter. Such connections produce positive rather than negative feedback. A special tertiary winding on an output transformer is necessary. A common-base stage with negative feedback is shown in Fig. 6-16. The feedback is voltage-derived and series-applied. Consequently the input impedance is raised, and the output impedance is lowered.

Feedback in the Common-Collector Circuit. The common-collector circuit is inherently a negative feedback amplifier. A little reflection will reveal this configuration as the end result of increasing current-derived series-applied feedback to the limit in a common-emitter stage. We must appreciate however that the shunting effect of a load connected across the emitter resistance decreases the transistor feedback current. In order to attain the high input impedance, the feature for which this circuit is often chosen, the load seen by the emitter resistor must not be too low.

GAIN CONTROL IN TRANSISTOR AMPLIFIERS

The control of gain in a transistor amplifier is preferably accomplished by varying some part of the circuit which affects the a-c level, but does not disturb the d-c operating point. This is particularly so in audio amplifiers. Figure 6-17 shows three methods of varying the gain of transistor amplifiers. In Fig. 6-17*A* the a-c input level to the second stage is varied by means of potentiometer R_1.

In Fig. 6-17*B* the current gain of the amplifier is governed by the amount of negative feedback selected by R_1. This method does not permit reduction of the transferred signal to zero level. However the attenuation range can be extended by provision of input and output circuits which see a greater impedance mismatch as feedback is increased. We note that the feedback is voltage-derived

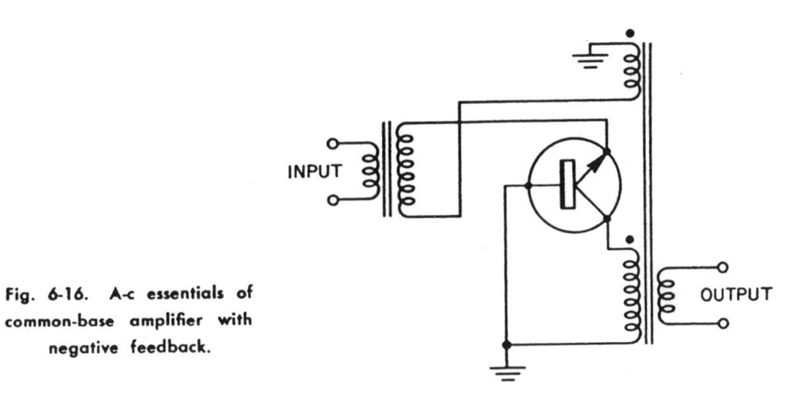

Fig. 6-16. A-c essentials of common-base amplifier with negative feedback.

shunt-applied. Accordingly, both input and output impedance decrease with increasing feedback.

In Fig. 6-17*C* the a-c essentials of a section of a superheterodyne receiver is depicted. The gain of the i-f amplifier is controlled by

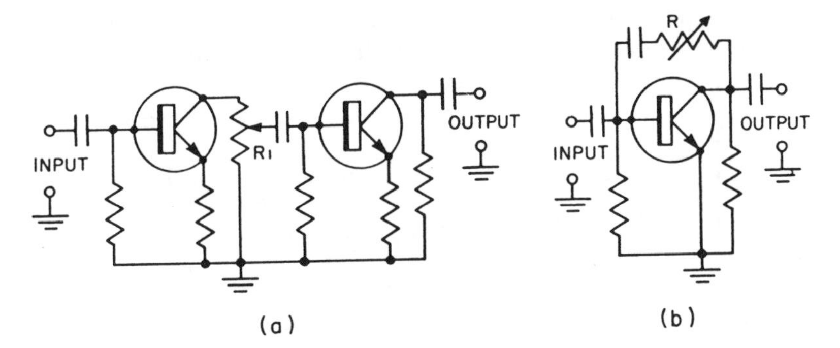

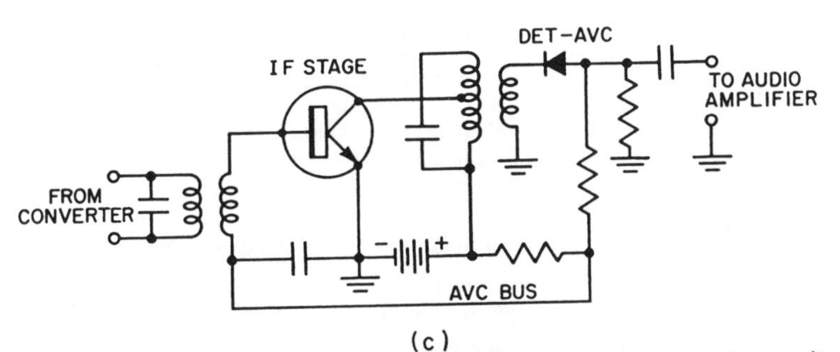

Fig. 6-17. Three gain-control methods: (A) Control of a-c input level to audio amplifier; (B) control of a-c gain by variable feedback in audio amplifier; (C) gain control by variation of base-emitter bias in i-f amplifier.

a d-c "error current" derived from the second detector. For strong signals, the forward bias applied to the base-emitter diode of the i-f amplifier is decreased. This lowers the current gain of the transistor. This effect exists only for very low values of emitter-base bias. In audio amplifiers, such operation would introduce considerable distortion due to nonlinearity of the transistor characteristics. In the i-f amplifier, "distortion" of the signal is of no consequence, because the harmonics and intermodulation products thereby generated are highly attenuated by resonant circuits.

TRANSISTOR NOISE

For many applications, the self-generated noise of the junction transistor is of no appreciable consequence. On the other hand, where a multiple-stage amplifier possessing a high gain-bandwidth product is part of the system, the internally produced noise power in the input stage can assume considerable if not limiting significance. At relatively low frequencies, generally including most

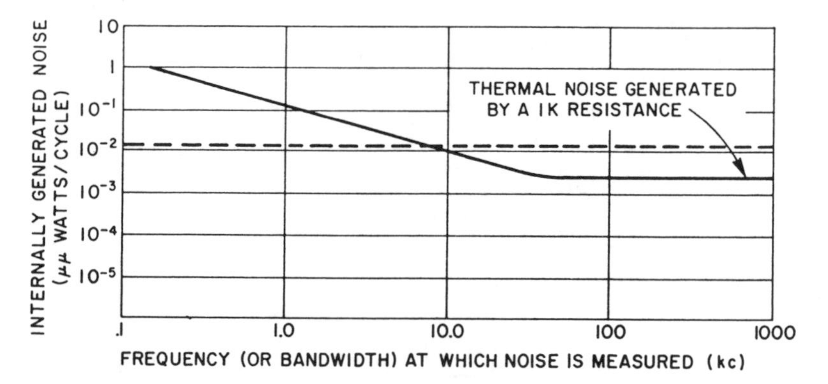

Fig. 6-18. Internally generated noise power as a function of frequency for typical small germanium junction transistor.

of the audio spectrum, the main source of transistor noise displays different frequency distribution properties than does the thermal, or "white noise," generated by resistance. The unique property of such transistor noise is that for a given bandwidth, it decreases in direct proportion to increasing frequency. Mathematically termed "$1/f$" noise, this relationship is equivalent to stating that the same noise power is contained in each octave (that is, each decade frequency range). At higher frequencies, a noise resembling shot noise in electron tubes assumes prominence (see Fig. 6-18). This noise, like thermal noise, has a "flat" power spectrum; i.e., this noise power increases with increasing octaves of the frequency spectrum. (We note that successively higher octaves of bandwidth contain more cycles per second and that noise can be assumed generated as so much power per cycle per second.)

Source of 1/f Noise. 1/f noise apparently has two sources, both however associated with the process of charge injection. One of these sources appears to be localized within the emitter region. The other emanates from the collector region. Low forward-bias current helps reduce the emitter noise power, whereas low collector voltage, say 1 volt or less, greatly diminishes collector noise power.

Other Noise Considerations of Junction Transistors. One of the advantages possessed by junction transistors over electron tubes is the absence of microphonics. This is, of course, due to the unitized physical structure of the transistor; no section is free to vibrate with respect to any other. Another advantage stems from the high immunity to induced noise from external magnetic or electric fields. In those critical applications wherein noise must be maintained at a low value, it is desirable to operate the input transistor stage from batteries rather than rectified and filtered alternating current. The input diode of the transistor responds to ripple as it would to a signal. In this respect, the vacuum tube may show up advantageously because the thermal inertia of a filament, or particularly an emissive cathode, makes such emitters less susceptible to emission modulation from moderate ripple percentages than the input diode of the transistor.

7. related semiconductor devices

Undeniably, the technologies of communications and control have benefited greatly from exploitation of the conventional triple-region junction transistor. We should appreciate, however, that this component is but a single member of a large family of semiconductor devices, each capitalizing upon a unique practical aspect of basic p-n junction phenomena. So numerous and varied are these, that the description in detail of even a reasonably representative portion of them could not possibly fall within the intended scope of this book. Nevertheless, our acquaintanceship with the junction transistor will certainly be enriched by giving at least some consideration to other semiconductor devices.

DOUBLE-BASE OR TETRODE TRANSISTOR

A structure which by its resemblance to the simple junction transistor, merits our first attention is the double-base, or tetrode, transistor. In its simplest form, we have an alloy junction transistor with a second base connection as depicted in Fig. 7-1. When the second base B_2 is biased with respect to the conventional base B_1, the emission of carriers from the emitter region is no longer approximately uniform, as in Fig. 7-2*A*. When B_2 is positive with respect to B_1, emission is considerably inhibited from the upper part of the emitter, as shown in Fig. 7-2*B*. Conversely, when B_2 is

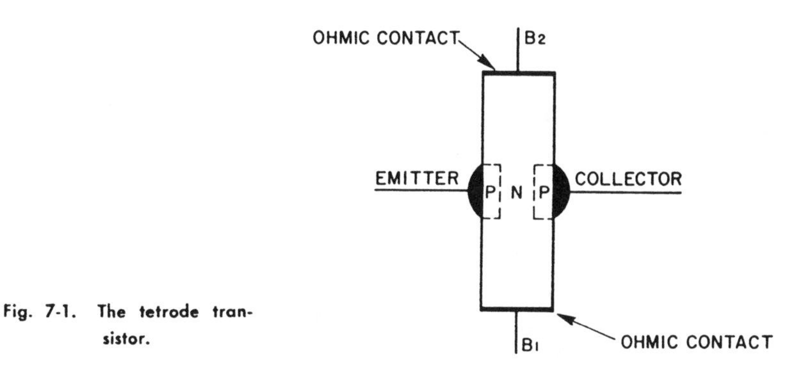

Fig. 7-1. The tetrode transistor.

negative with respect to B_1, the preponderance of emission occurs from the upper portion of the emitter (see Fig. 7-2*C*).

In either case, the effective areas of both emitter and collector are considerably reduced. This in itself contributes significantly to improved high-frequency performance. Of even more importance at high frequencies the effective base resistance is greatly reduced because less base material is involved in carrier transport between emitter and collector. (We note this effect is opposite to that which would be expected from ordinary experience with metallic conduction.) Figure 7-3 shows an i-f stage using a tetrode transistor.

As the bias on B_2 is increased, the α of the transistor will decrease because an appreciable portion of emitted carriers will be deflected as they are about to enter the collector junction. Such deflected carriers recombine in either the B_1 or B_2 region and thus

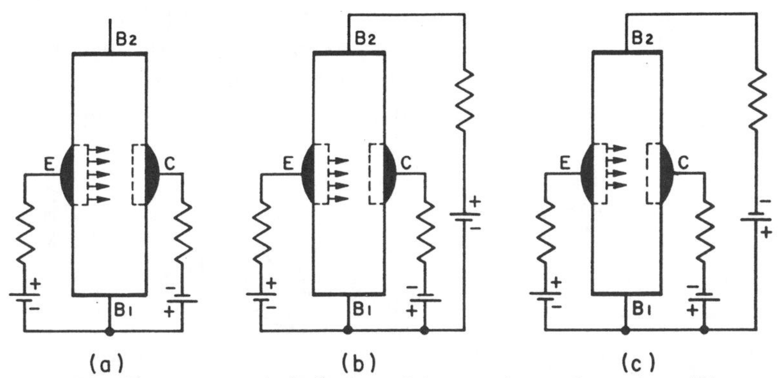

Fig. 7-2. Electronic control of effective emitting area in tetrode transistor: (A) no bias on B_2; (B) B_2 is negative with respective to B_1; (C) B_2 is negative with respect to B1.

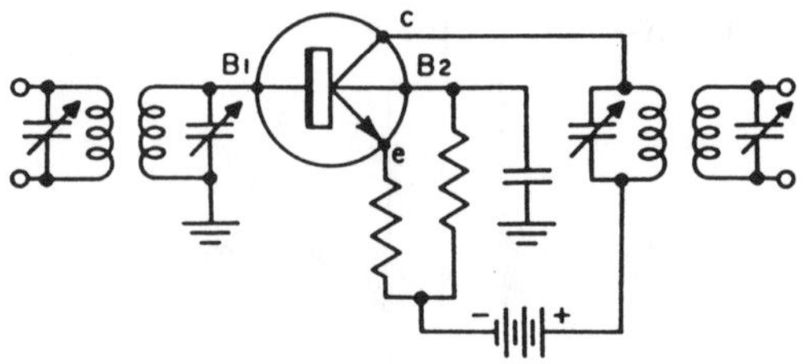

Fig. 7-3. Tetrode transistor i-f amplifier.

do not participate in transistor action. We see that the control of bias applied to B_2 affords an excellent means of controlling circuit gain. This is particularly so inasmuch as signal circuits are not directly affected. We are reminded somewhat of screen grid control of a-c gain in a tube amplifier.

THE HOOK TRANSISTOR

The hook transistor makes use of a collector region which contains an extra p-n junction. This structure is illustrated in Fig. 7-4. We recall that in a conventional three-region junction transistor a current amplification of $1/(1-\alpha)$ exists in the base lead relative to the emitter lead. It is this very multiplying factor which enables the common-emitter circuit (base injection) to produce more current gain than the common-base circuit (emitter injection). Thus, the ratio of β to α is $1/(1-\alpha)$. The incorporation of an additional junction in the collector region of the hook transistor provides such current multiplication.

Further insight into the nature of current multiplication may be attained by inspection of the approximately equivalent circuit shown in Fig. 7-4*b*. Here we see a tandem arrangement consisting of a common-base amplifier driving a common collector amplifier.

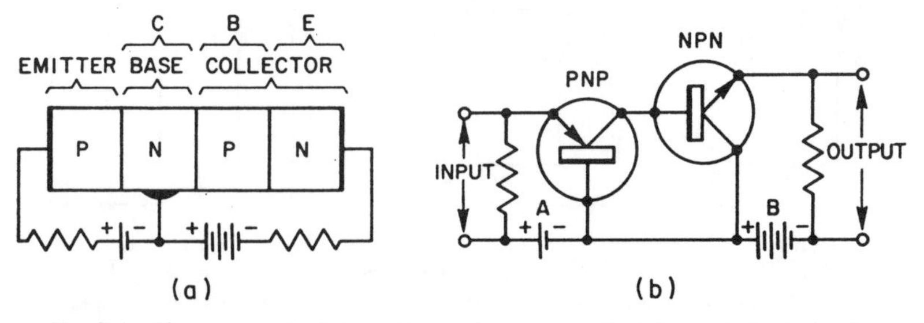

Fig. 7-4. The p-n-p-n hook transistor and an approximately equivalent circuit.

Complementary symmetry takes care of d-c bias requirements. Voltage gain is provided by the input stage, current gain is produced by the output stage. We note that emitter-base forward bias to the n-p-n transistor as well as reverse bias for the collector-base diode of the p-n-p transistor is supplied by battery B. Referring back to Fig. 7-14*A* the regions designated emitter, base, and collector may be correlated with the respective elements of the p-n-p stage of the circuit shown in Fig. 7-4*B*. The regions designated *C, B,* and *E* are by the same reasoning the collector, base, and emitter of the n-p-n output stage. Thus, three of the four regions of the p-n-p-n structure can be construed to perform double functions in terms of this analysis.

Unlike conventional junction transistors, the hook transistor produces an α in excess of unity when connected as a common-base amplifier. This endows it with a negative input impedance

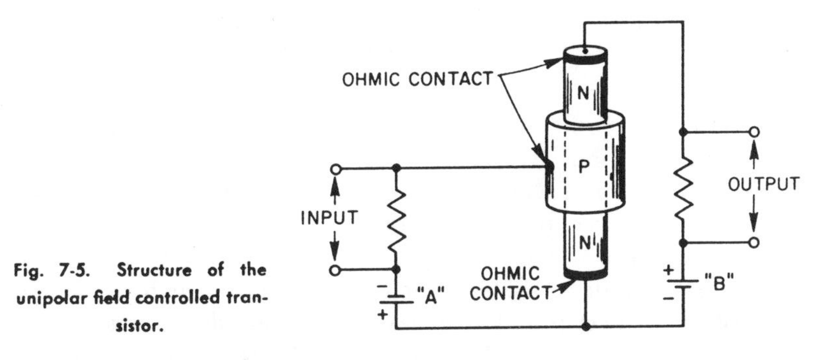

Fig. 7-5. Structure of the unipolar field controlled transistor.

which is useful for certain switching applications. A particularly significant characteristic of the hook-transistor is that it suggests the mechanism which may operate in point-contact transistors which also display common-base current gain factors in excess of unity.

THE UNIPOLAR FIELD CONTROLLED TRANSISTOR

The unipolar field controlled transistor is an amplifying device in which semiconductor phenomena are used differently than in the conventional transistor. In Fig. 7-5 the n-type germanium rod is continuous from end to end. The central region of this rod is encircled by a band of p-type germanium, and a p-n junction exists between the band and the rod. We see that this structure does not consist of alternate conductivity regions in the same way as conventional transistors are constructed. The effective cross-sectional area of the section of rod encircled by the band is a function of the width of the depletion layer which separates the p- and n-type material. This is illustrated in Fig. 7-6. The width of the

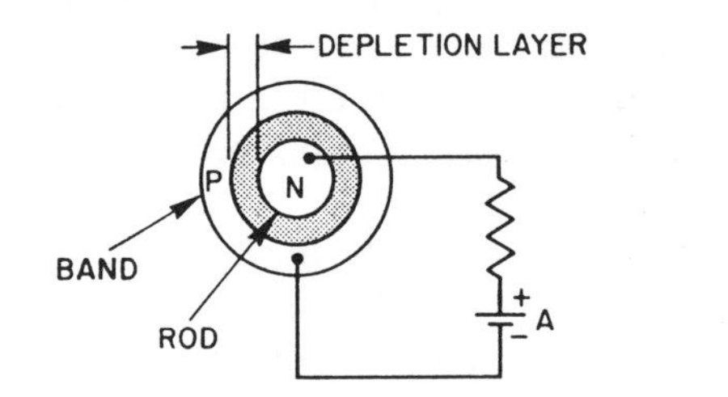

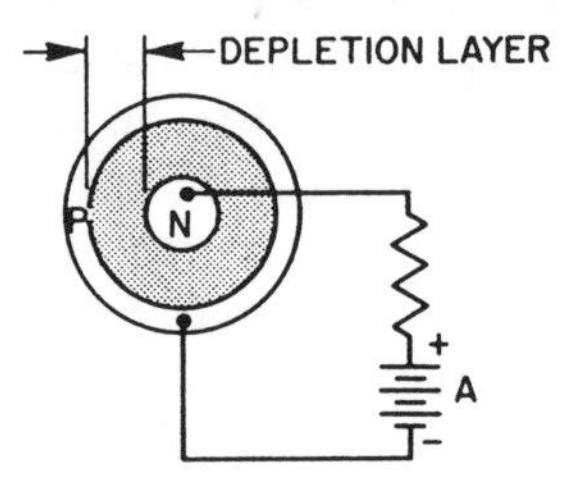

Fig. 7-6. Electronic control of effective cross-sectional area in rod of unipolar field transistor.

depletion layer is, in turn, governed by the reverse bias impressed across the junction. By varying this bias, we modulate the conductivity of the rod. Consequently a current passing through the rod will reproduce these variations. Power amplification results because, whereas a minute current is consumed by the reverse biased input diode, considerable current passes through the n-type rod. We note that this device has no "output diode," as does a conventional transistor. Therefore, the polarity of the rod supply is of little consequence. Battery B could be reversed if desired.

THE DOUBLE-BASE DIODE

The double-base diode, or unijunction transistor exhibits properties which make it suitable for switching operations rather than for proportional amplification. The structural configuration of this device is shown in Fig. 7-7. Two ohmic contacts, B_1 and B_2, are made to the end sections of an n-type germanium bar. In the central region the bar is a p-n junction with a lead brought out from the pellet of p material. Suppose a d-c voltage with polarity as indicated is impressed across B_1 and B_2. The current that flows longitudinally through the germanium bar will cause the central region of the bar to be positive with respect to base B_1. As a consequence, a small positive voltage impressed between E and B_2 will not suffice to make E positive with respect to the germanium bar. Therefore, such a small voltage will not cause forward conduction in the p-n junction.

We should note furthermore that the lower portion of the emitter region (that which is closest to B_2) can be expected to emit before the middle or upper portions, inasmuch as the latter portions are adjacent to more highly positive germanium. In essence the middle and upper portions of the emitter are back-biased by a higher voltage than the lower portion.

If we slowly increase the positive emitter voltage, a point will be reached wherein the lowest portion of the emitter becomes positive with respect to adjacent germanium, and forward conduction occurs. The advent of forward conduction upsets the longitudinal voltage distribution along the germanium bar in such a manner that more of the emitter region finds itself positive with respect to adjacent germanium and heavier forward conduction ensues. This phenomenon is cumulative and rapidly propagates to embrace the entire emitter region.

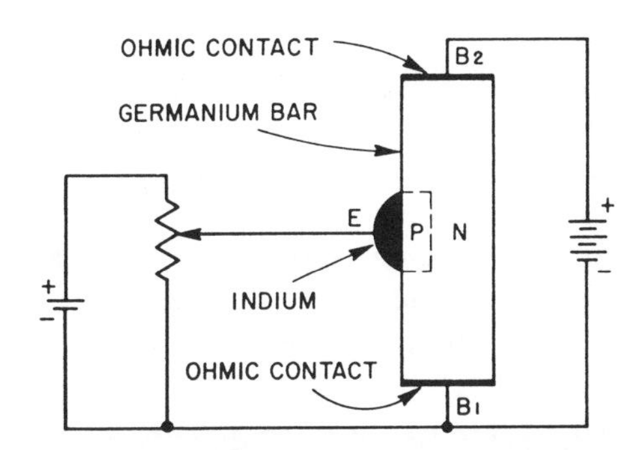

Fig. 7-7. The structure of the double-base diode and its biasing arrangement.

We see that a switching cycle has taken place; a tiny forward current initiates the occurrence of heavy conduction in the E-B_1 circuit. The process is essentially one of conductivity modulation in the lower half of the germanium bar brought about by emission of holes into this section of the bar.

A self-actuating switching circuit, or sawtooth generator, is shown in Fig. 7-8. When this circuit is placed into operation by closing switch S, the capacitor begins to charge from current flowing through the p-n junction. We note that the junction is reverse-biased, E being negative with respect to B_2. As the voltage across the capacitor builds up, a value is finally attained which permits the lower portion of the emitter region to become positive relative to adjacent n-type germanium. When this occurs, the entire junc-

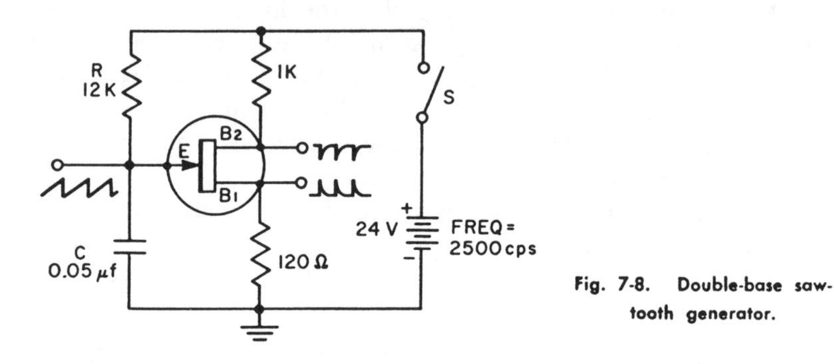

Fig. 7-8. Double-base sawtooth generator.

tion rapidly becomes forward-biased due to abrupt conductivity change in the lower half of the germanium bar. The capacitor is thereby discharged to a low level and the entire cycle of events then repeats itself. The 12-k resistance provides a slight forward bias to the emitter in order to make oscillation self-starting. This circuit is readily synchronized as a stable frequency divider by injecting trigger pulses as indicated in Fig. 7-9.

THE SILICON CONTROLLED RECTIFIER

The silicon controlled rectifier, like the hook transistor, comprises four alternately doped conduction regions. This device may be said to be the semiconductor counterpart of the thyratron gas tube. As with the thyratron, conduction is initiated by a trigger signal which must attain a certain level. Likewise, in thyratron fashion, once conduction exists, the triggering signal exerts no further control unless the load circuit is interrupted. The silicon controlled rectifier possesses the ability to switch very heavy currents in a matter of microseconds.

The basic structure of this device together with an approximately equivalent circuit is shown in Fig. 7-10. Except for the insignificant fact that we happen to be now considering an n-p-n-p

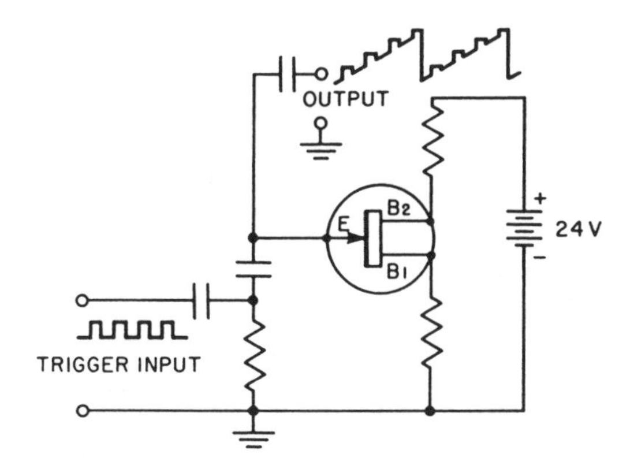

Fig. 7-9. Double-base diode connected as a synchronized frequency divider.

rather than p-n-p-n structure, we see that the silicon controlled rectifier is essentially of the same nature as the hook transistor. The essential difference is that the silicon controlled rectifier is designed to function as a large signal-switching device, whereas the

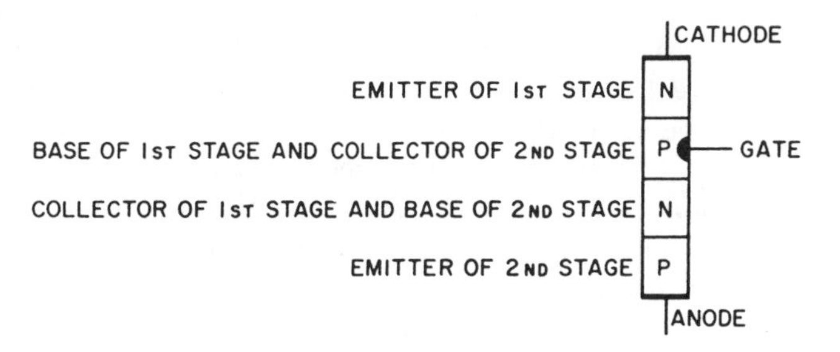

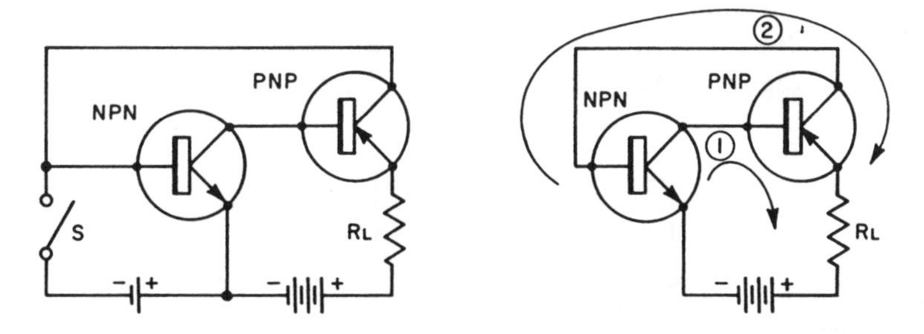

Fig. 7-10. The silicon controlled rectifier: (A) basic structure; (B) its approximately equivalent circuit; (C) flow of load current after conduction has been initiated by momentarily closing switch *S*.

hook transistor is intended to provide proportionate amplification of small signals. We note that the two-transistor analogy in Fig. 7-10*B* resembled a multivibrator circuit in that the collector of each transistor is returned to the emitter of the alternate transistor. In this circuit, sustained oscillation does not exist because both

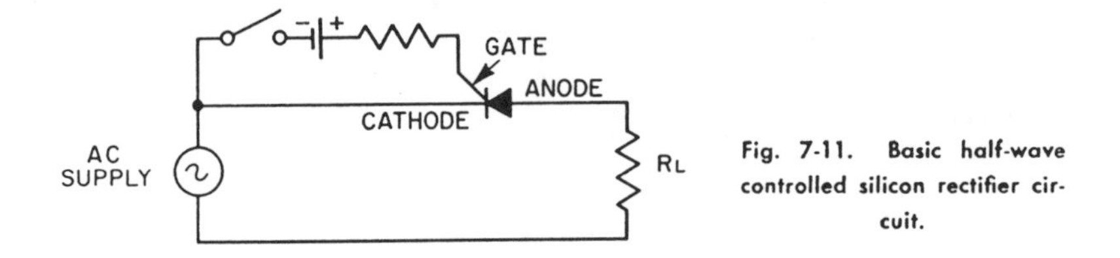

Fig. 7-11. Basic half-wave controlled silicon rectifier circuit.

transistors are deprived of forward bias when they are in their OFF conduction states, and there are, furthermore, no energy storage elements, such as capacitors, to provide time constants for the relaxation-type oscillations associated with multivibrators. Initially, the circuit is in its OFF condition, and no current flows through the load. This is because the emitter-base diode of the second stage does not receive forward conduction bias unless the first stage is in its conductive state. However, the first stage is nonconductive because switch *S* is open. We now momentarily close switch *S*, establishing conduction in the first stage and thereby in the second stage also. Two paths now exist for the load current as indicated in Fig. 7-10*C*. Path 1 maintains forward conduction bias on the second stage, whereas path 2 maintains forward conduction bias on the first stage. The currents in the two paths reinforce one another, and the conductivity of both transistors is well in their saturation regions.

A half-wave controlled rectifier circuit is shown in Fig. 7-11, whereas the full-wave arrangement is depicted in Fig. 7-12. In

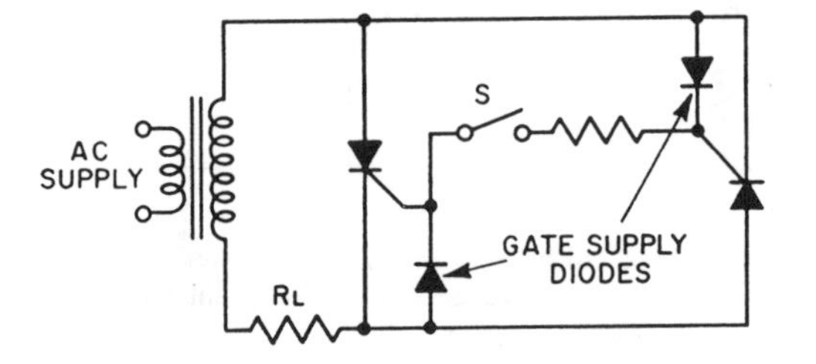

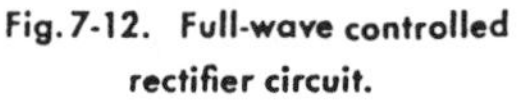
Fig. 7-12. Full-wave controlled rectifier circuit.

both cases a very small actuating current suffices to turn on or off the flow of heavy load currents. We note that the gate signal must remain present for continued flow of load current. This is because the current through the controlled rectifiers is ac and conduction ceases for approximately half of the applied cycle.

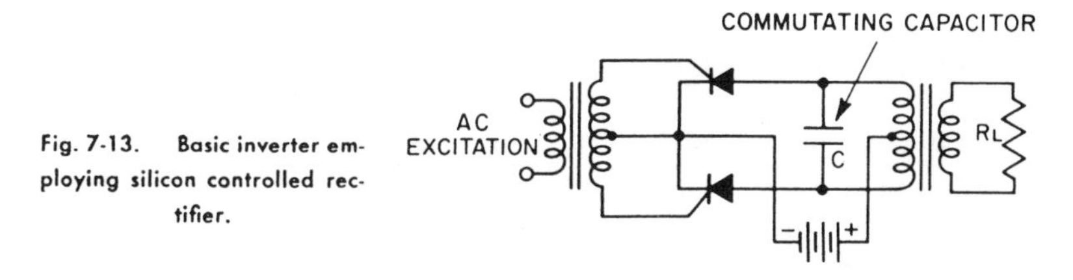

Fig. 7-13. Basic inverter employing silicon controlled rectifier.

When the polarity of the current reverts to that necessary for conduction, the gate signal must be present to cause renewed firing. Instead of the switch, a control device such as a transistor may be used to satisfy a particular control requirement.

Figure 7-13 illustrates the essentials of an inverter. Capacitor *C* serves the same function as in similar thyration inverters. When one controlled rectifier is switched to its conductive state, the resultant pulse communicated through this capacitor momentarily depresses the voltage existing across the alternate rectifier, thereby switching it to its OFF condition. The output is a square wave with a frequency of the excitation applied to the gates. The waveshape and duty cycle of the exciting current is of little consequence because the switching operation is mainly governed by the commutating capacitor in conjunction with the characteristics of the controlled rectifiers.

Further insight into the characteristics of the silicon controlled rectifier can be gained by inspection of Fig. 7-14. It is seen that if moderate a-c voltage is applied, say not exceeding an amplitude corresponding to the range between points 1 and 2, negligible current flows through the device. Under such a condition, the rectifier behaves as an open switch to both halves of the a-c cycle. If the amplitude is increased in the forward direction much beyond point 2, sufficient transistor action will occur to regeneratively switch operation to the *AB* region in which the device acts as a closed switch for one half of the a-c cycle.

Significantly, however, operation can be shifted to the *AB* region *without* raising the applied voltage beyond point 2. This is, of course, accomplished by means of a momentary gate signal (see Fig. 7-15). The gate signal will switch operation to the *AB* region for any forward bias exceeding about 0.75 volt but less than the breakdown value indicated by point 3. The reverse breakdown region is of little practical concern other than that such operation should be avoided.

THE FOUR-LAYER DIODE

The four-layer diode is a semiconductor switching device utilizing much the same principles as the silicon controlled rectifier. The four-layer diode is a two-terminal device and is made in smaller power ratings than the three-terminal silicon controlled rectifier. In our discussion of the silicon controlled rectifier, it was noted that switching could be initiated by the application of

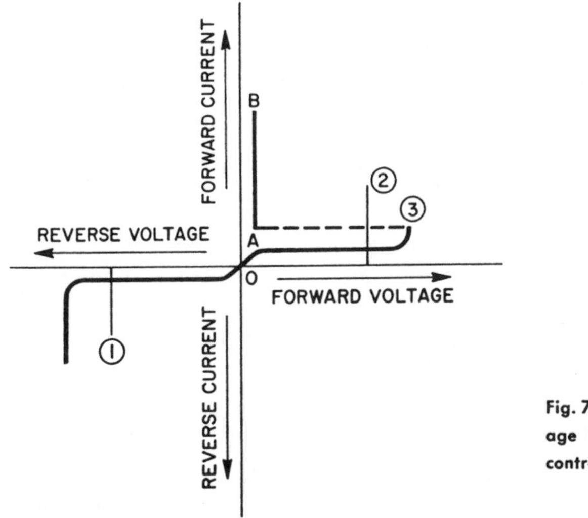

Fig. 7-14. Effect of a-c voltage impressed across silicon controlled rectifier with gate circuit open.

sufficient forward voltage even with the gate circuit open. In the case of the four-layer diode, this switching technique is intentionally used.

Referring back to Fig. 7-10*B*, we note that the two-transistor analogy constitutes a closed positive-feedback arrangement. In such systems, either sustained oscillation or a single switching cycle ensues when the over-all loop gain is equal to unity. Such a condition would exist when the α of each transistor is equal to ½. Inasmuch as no forward bias is applied to the emitter-base diodes of the transistors, the α's are considerably less than the critical value of ½ when a moderate forward voltage is applied across the device. It is important to appreciate that, due to the leakage current I_{CO} of each transistor, a tiny current will flow through the device. This is because the I_{CO} of each transistor provides a small forward bias to the alternate transistor. Thus, for moderate forward potentials impressed across the terminals of the four-layer diode, some transistor action will occur. Inasmuch as the current gain of a silicon transistor is low for small forward-bias currents, the circuit will lack sufficient gain to regeneratively switch conductive states. However, if we increase the forward potential, we will arrive at a point where the collector-base junction of the p-n-p transistor approaches avalanche breakdown. This immediately increases the current passing through the terminals of the circuit. The increase in forward-bias current at the emitter-base junctions of both transistors increases their current gains, which thereby enables a further increase in current through the circuit terminals. At the same time the initial avalanching of carriers in the p-n-p transistor rapidly approaches saturation.

We see that the factors participating in the switching process are cumulative. The entire process is speeded by the regenerative

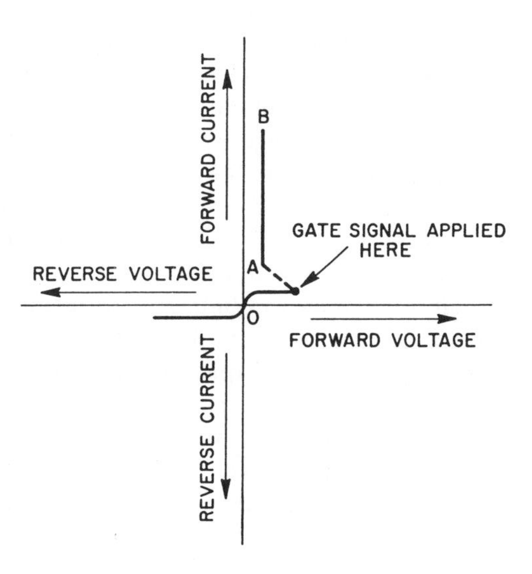

Fig. 7-15. Effect of gate signal on forward conduction of silicon controlled rectifier.

action of the circuit, so that once initiated, closure of our semiconductor switch occurs in a very brief time, on the order of perhaps 10 to 100 mμsec. Once closed, a small "holding" current suffices to maintain the transistors in their highly conductive

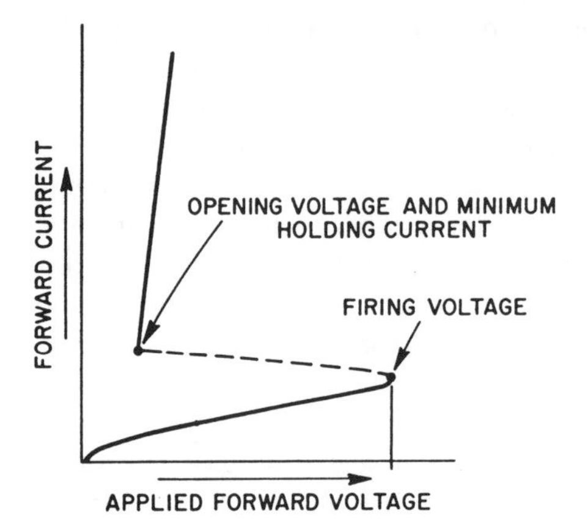

Fig. 7-16. Forward breakdown characteristic of the four-layer diode.

states, each transistor reinforcing its alternate in conduction. If the current is decreased from this threshold value, the drop in current gain in the two transistors causes the circuit to revert to its nonconductive or OFF state.

The four-layer diode is, as a circuit element, somewhat analogous to the simple neon lamp and can be connected in a relaxation oscillator circuit which charges and discharges a capacitor in similar fashion to the well-known neon bulb oscillator. In both devices, the ability to behave as a self-driven switch derives from a voltage hysteresis; that is, firing voltage is greater than opening voltage. This is shown in Fig. 7-16. We see in Fig. 7-17 a four-layer diode connected as a relaxation oscillator. The series resistor R_g limits the current to a value below that required for maintaining the four-layer diode in its closed condition. The capacitor charges through R_g until sufficient voltage difference exists across its plates to fire the four-layer diode. The four-layer diode discharges the capacitor, then reverts to its open condition, whereupon the capacitor-charging cycle repeats.

In Fig. 7-18 there is shown a pulse generator which is triggered from an external source. The circuit is essentially similar to that of the relaxation oscillator. However, in this case the d-c voltage applied across the terminals of the four-layer diode is less than breakdown voltage. An incoming waveform of sufficient amplitude then initiates breakdown in the four-layer diode, and the resultant pulse derives its power from the d-c source. The performance of this circuit reminds us of the Schmitt trigger which is widely used for conversion of slow waveforms to pulses with steep sides.

The four-layer diode is a versatile switching element and is readily adaptable to circuits requiring rapid transition in operating state. Such applications include binary and ring counters, and the various pulse techniques underlying the operation of digital apparatus.

THE DYNISTOR

The dynistor is a device similar to the four-layer diode. The parent material is, however, germanium rather than silicon. The structure, as indicated in Fig. 7-19 is p-n-p-m signifying the use of a metallic region as cathode. This is possible because a metal

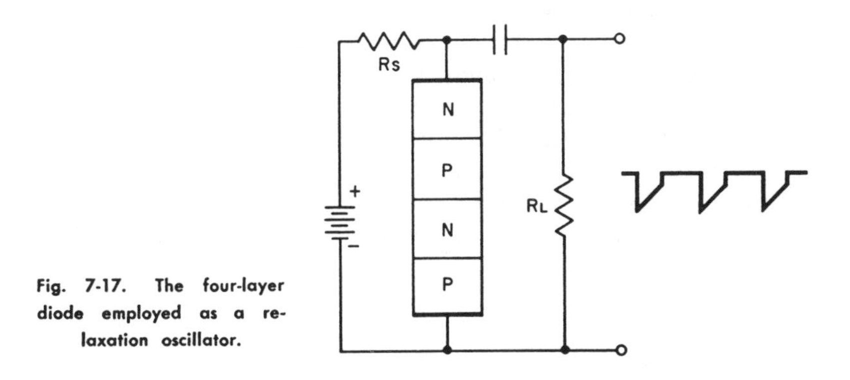

Fig. 7-17. The four-layer diode employed as a relaxation oscillator.

can be used as a source of electrons, somewhat as n-type semiconductor material. However, when a metal is used as an emitter, a true p-n junction is not obtained. For reverse conduction, the metal acts more in the nature of an ohmic contact. Consequently, the dynistor conducts for reverse polarity of the applied voltage in contrast to the four-layer diode which blocks reverse current flow. The salient feature of the dynistor is its exceedingly low turn on time, on the order of 0.01 μsec.

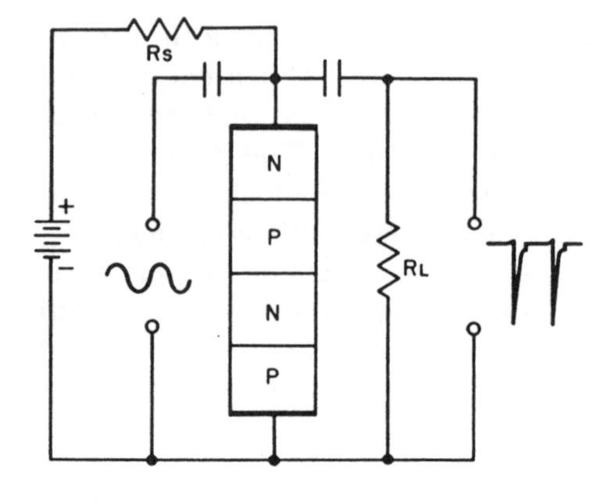

Fig. 7-18. The four-layer diode connected in a pulse generator circuit.

BILATERAL TRANSISTOR

If we exchange emitter and collector leads in a common-base amplifier using an ordinary junction transistor, it is often possible to retain an appreciable amount of power gain; that is, the transistor is inherently a bilateral device, but its design is such as to favor one direction of signal transmission over the other. Thus, a transistor may have an α of 0.970 in its intended forward direction and an α of 0.800 if used with emitter and collector leads interchanged. The corresponding β's are 32 and 4.

Obviously degradation in performance would be much worse in the common-emitter than the common-base circuit. Favored direction of amplification is brought about primarily by the use of different doping and different effective areas in emitter and collector regions. However, in the bilateral transistor of Fig. 7-20

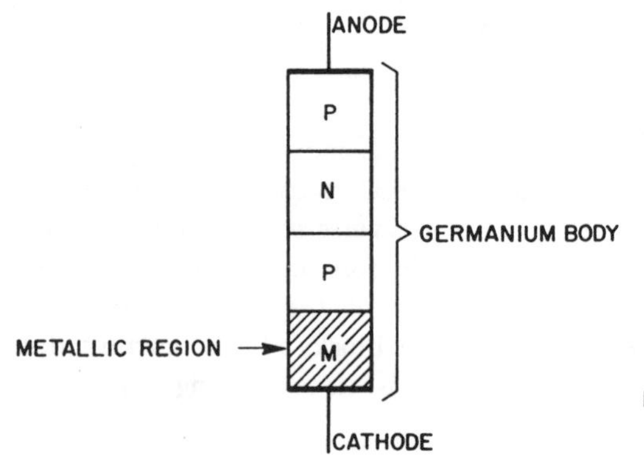

Fig. 7-19. Structure of the Westinghouse dynistor.

the emitter and collector regions are identical, making possible bilateral signal transmission as well as the interchange of emitter and collector leads. This is a desirable feature for carrier telephone systems, phase-detection circuits, bidirectional switching, and multiplexing devices. It should be appreciated that bilateral transmission is not simultaneous. Rather, it is necessary to exchange emitter and collector bias supplies to reverse the direction of transmission.

THE ZENER DIODE

The Zener diode makes use of both Zener and avalanche breakdown which occur when the reverse voltage impressed across a p-n junction exceeds a certain value. The semiconductor material used is silicon.

Although reverse voltage breakdown exists in germanium junction diodes, a much more abrupt transition is obtained when silicon is employed. A series current-limiting resistance prevents the breakdown from being destructive. Figure 7-21 depicts the characteristics of a typical Zener diode. Prior to breakdown voltage, the reverse current through the diode is very small by virtue of the small number of carriers thermally generated in the depletion layer. After breakdown occurs, the resultant multiplication of carriers enables a very heavy flow of current.

We are reminded of the behavior of a gaseous diode subjected to an ionization voltage. The Zener diode is used in similar applications as the gaseous diode, but with definite advantages in performance. Unlike the gaseous diode, a wide range of breakdown voltages are available, from a few volts to 150 volts or more.

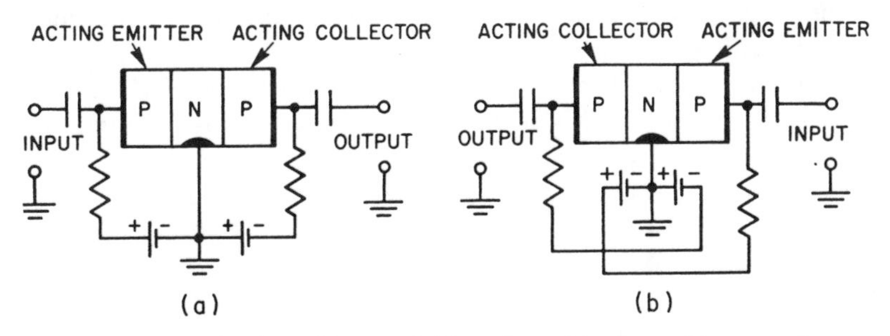

Fig. 7-20. Example of possible use of bilateral transistor in an intercommunications amplifier: (A) bias connected for "listen"; (B) bias connected for "talk."

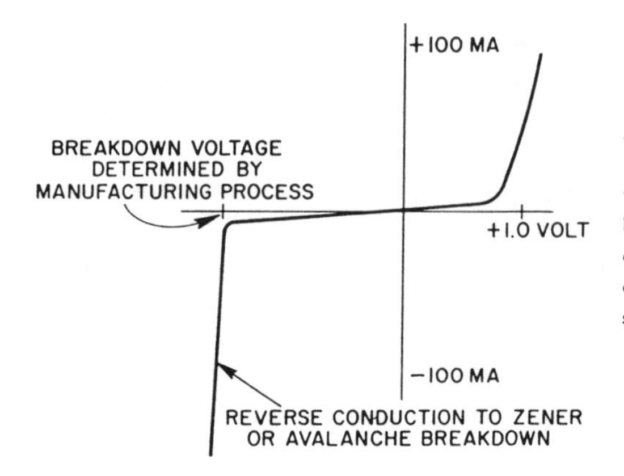

Fig. 7-21. Characteristic of the Zener diode. Reverse breakdown is caused by Zener phenomena when the breakdown voltage is less than about 5½ volts. For higher breakdown voltages, carrier avalanche is the predominant mechanics responsible for the abrupt increase in conduction.

As in the gaseous diode, the breakdown phenomena are accompanied by a nearly constant voltage drop across the device. Consequently, the Zener diode makes a very versatile voltage-regulating element. It does not require a firing voltage in excess of operating voltage as does the gaseous diode.

The basic circuits for regulation of d-c and a-c voltages are shown in Fig. 7-22. In the a-c arrangement, we see two Zener diodes series-connected back-to-back. This circuit functions also as a peak clipper and will provide a good approximation to a square wave if the amplitude of the input sine wave is high. The two diodes are alternately switched from reverse conduction to forward-conduction states. For the portion of a half cycle during which one diode regulates, the other operates as a closed switch.

An element such as the Zener diode which maintains near-constant terminal voltage over wide variations of current can provide a similar circuit function to that of a large bypass capacitor. Figure 7-23 shows a pair of Zener diodes connected to produce cathode-grid bias in a vacuum-tube amplifier. No bypass capacitor is necessary inasmuch as the a-c dynamic impedance of the Zener diodes is very low. This is very convenient for low frequencies

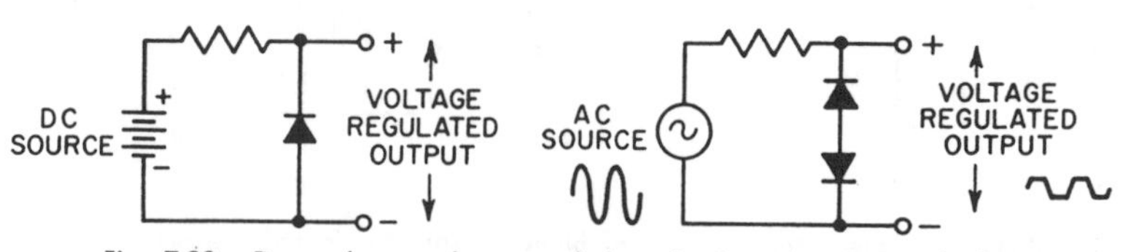

Fig. 7-22. D-c and a-c voltage-regulating circuits using Zener diodes.

where the conventional cathode resistor would require an inordinately large bypass capacitor in order to avoid undesired degeneration.

A further advantage of the Zener diode bias arrangement is that the a-c impedance of this bias source is not frequency-dependent as is a capacitor.

An important use of the Zener diode is as a precise voltage reference in regulated power supplies and other servo systems. In these applications, the Zener diode substitutes for a battery or a gaseous voltage-regulator tube (see Fig. 7-24). Yet another function is the automatic protection of meter movements or other sensitive devices from overload. Figure 7-25 shows how such protection is provided for the 100 μamp movement of a 10-volt d-c meter. The Zener diode is rated at 8 volts breakdown. When a voltage in excess of 10 volts is applied across the 10-volt terminals,

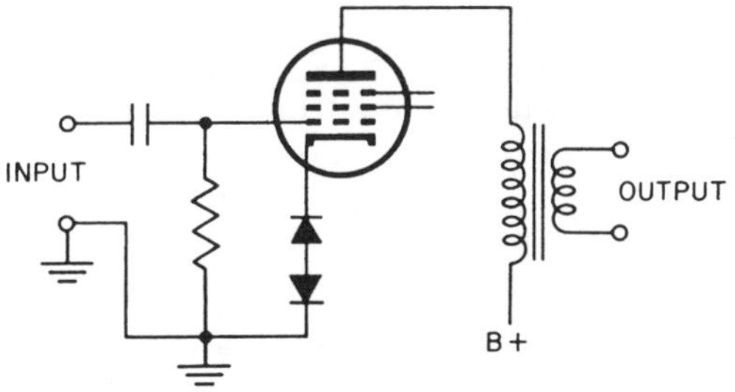

Fig. 7-23. Zener diode employed as source of cathode-grid bias in vacuum-tube amplifiers.

the Zener diode breaks down, thus preventing more than full deflection current from flowing through the meter movement.

It is interesting to note that there is no voltage hysteresis associated with reverse breakdown. This precludes use of the Zener diode as a relaxation oscillator such as is obtained by connecting a resistance in series with the parallel combination of a neon bulb and a capacitor.

THE SEMICONDUCTOR VARIABLE CAPACITOR

The semiconductor variable capacitor is a silicon junction diode designed to advantageously exploit the voltage dependent width of the depletion layer. The "plates" of the capacitor may be said to constitute the edges of the p and n regions adjacent

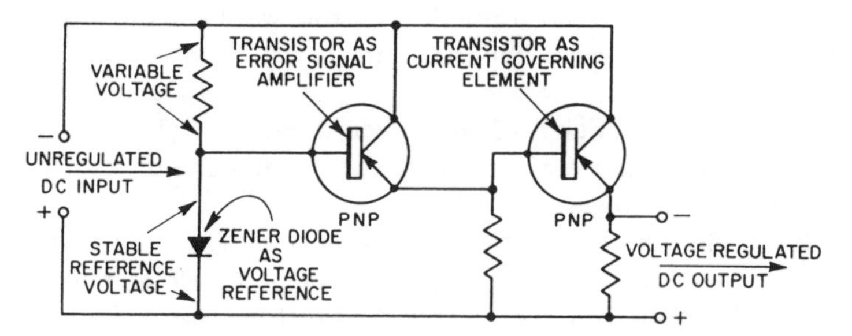

Fig. 7-24. The Zener diode as a voltage reference in a regulated d-c power supply.

to the depletion layer (see Fig. 7-26). The dielectric of the capacitor is the near-intrinsic material in the depletion layer itself. When the reverse bias impressed across such a diode is increased, the depletion layer widens, thereby reducing the effective capacitance. This effect is very useful for electrically tuning resonant circuits, particularly when automatic frequency correction is desired in response to a change in a d-c error signal. Another interesting use is found in the dielectric amplifier. Here, a change in resonant frequency along the steep slope of a high-Q tank circuit results in magnification of the bias change responsible for the frequency shift.

It should be mentioned here that there is another type of voltage-sensitive variable capacitor involving the material barium titanate in polycrystalline form. In this device the *dielectric constant* of the material changes with the voltage impressed across the plates. This type of voltage-tuned capacitor is considerably more temperature-sensitive than is the junction diode type.

THE JUNCTION PHOTODIODE

The junction photodiode is a light-actuated p-n device. In the ordinary junction diode, the current which flows under reverse-bias condition is due in large part to the thermally gen-

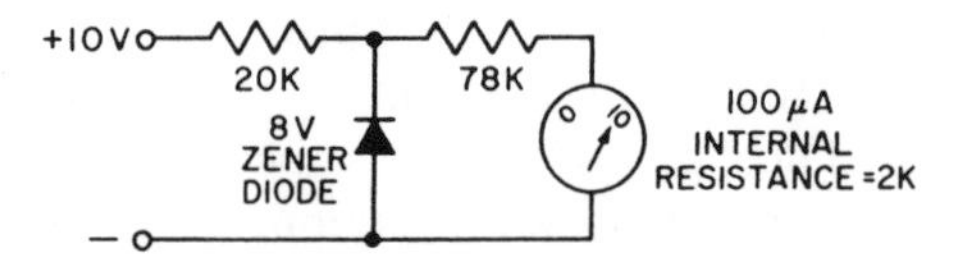

Fig. 7-25. The Zener diode employed as a protective device for a sensitive meter.

erated electron-hole pairs in the depletion layer. Light is basically the same physical manifestation as heat, but occupies a higher frequency band in the electromagnetic spectrum. Both heat and light manifest corpuscular as well as wave characteristics and can thereby be thought of as traversing space and materials in clusters of discrete wavetrains or packages of energy known as *photons.*

If the junction region of a diode is exposed, light photons will impinge on the atoms in the semiconductor material. The impact energy will be sufficient to dislodge some outer-orbit electrons associated in the covalent bond which unites the atoms in the crystal-lattice structure. The dislodged electrons become mobile current carriers and simultaneously produce holes which likewise

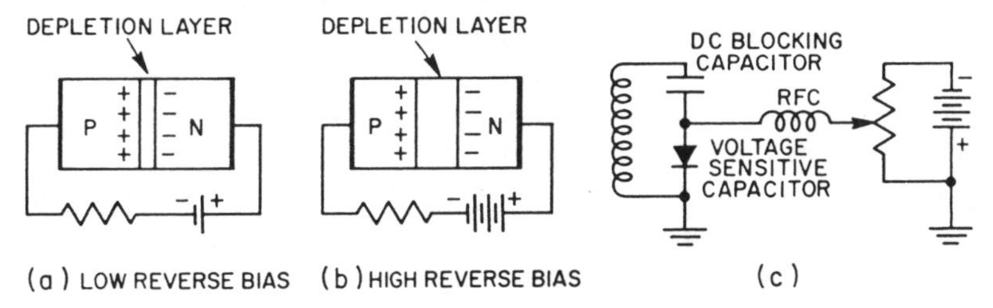

Fig. 7-26. The voltage-sensitive capacitor: (A) junction is narrow, producing a relatively high effective capacitance; (B) junction is wide, producing a relatively low effective capacitance; (C) turning a resonant circuit by means of a variable d-c voltage.

behave as mobile current carriers. These free electrons and holes are swept out of the junction region by the electric field produced by the bias battery and give rise to current in the external circuit (see Fig. 7-27).

THE SOLAR GENERATOR

The solar generator (see Fig. 7-28) is a p-n junction diode in which an entire surface of the depletion layer is permitted to receive light photons from solar radiation. The diode, or cell, consists of a thin slice of arsenic-doped silicon into which has been diffused a microscopically thin layer of boron. The major portion of the silicon wafer is thus of the p-conductivity type, whereas the surface is of the n variety. Ohmic contacts are made to these p-n junction elements. The surface layer of n material is so thin

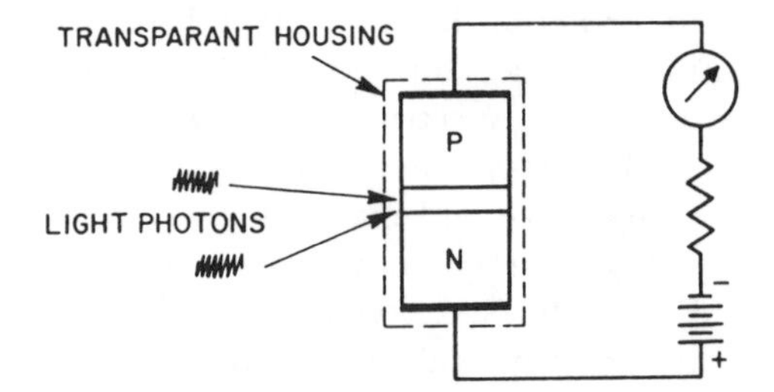

Fig. 7-27. The junction photodiode.

that a considerable fraction of the sun's radiation impinging at the surface penetrates through to the depletion layer of the junction.

Here, the resultant action is the same as in the photodiode. Electron-hole pairs are generated which upset the electrical equilibrium of the p-n junction, thereby producing a current in the external circuit. The efficiency of the silicon solar generator is about 11%, considerably in excess of other thermoelectric generators such as the thermocouple.

With this efficiency, over 10½ watts/ft² of silicon surface is available from strong sunlight. The open-circuit voltage of a single cell is in the neighborhood of 5½ volts. Under matched load conditions the output voltage drops to about 0.4 volt. Output current generally ranges from several to 150 ma. The cells may be connected in series, parallel, or series-parallel in order to provide a desired operating voltage and current. Such combinations are often referred to as *solar batteries*. However, this term may be somewhat misleading, inasmuch as energy cannot be stored as in the case of electrochemical cells or batteries. Indeed, a useful function is provided by employing chemical-storage batteries in conjunction with the solar cells. Surplus electrical energy is then stored as chemical energy, which in turn is immediately available as electrical energy at night and other times when solar radiation is not available.

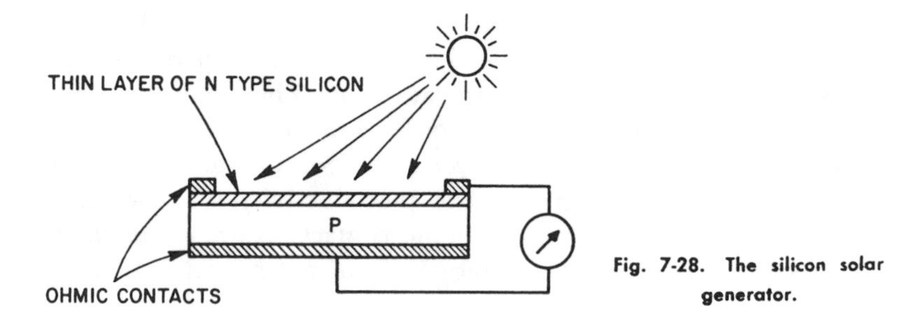

Fig. 7-28. The silicon solar generator.

THE PHOTOTRANSISTOR

The phototransistor possesses an advantage over the photodiode in that power amplification is inherently developed. In the simplest usage of the phototransistor, unmodulated light is permitted to impinge upon an exposed edge of the emitter-base junction region. The base lead is left unconnected. The arrangement resembles that of the photodiode inasmuch as only two leads are connected to the external circuit. However, true transistor action takes place because forward emitter-bias is produced by the release of electron-hole pairs by impacting light photons. The connections for operation of the phototransistor from an unmodulated light source are shown in Fig. 7-29*A*. When the light source is modulated, it is advantageous to provide a slight forward bias to the emitter-base diode, as shown in Fig. 7-29*B*.

Figure 7-30 depicts a more practical circuit for modulated light. A resonant circuit is employed in order to maximize the signal-to-noise ratio. Here again the input diode is operated with a small forward bias. We see that for nonmodulated light the phototransistor behaves as a d-c amplifier, whereas a-c amplifier action is obtained for modulated light. The spectral response of both the junction photodiode and junction phototransistor embraces the visible region and generally extends far into the infra red.

THE TUNNEL DIODE

The latest, and most spectacular, semiconductor device (at the time of writing of this book) is the *tunnel diode*. The theory underlying the operation of the tunnel diode is quite complex, involving speculative hypothesis in quantum mechanics. Nevertheless, the physical configuration, as well as associated circuitry required, is simpler than that of the transistor. In many respects, both attained and predicted performance are superior to the best available transistors.

The tunnel diode is a two terminal device, somewhat similar to the p-n junction diode commonly employed as a rectifier, or voltage regulator. The significant difference is that the tunnel diode displays a negative resistance region throughout a portion of its forward-conduction region. (This is in contrast to the negative resistance region exhibited by some point-contact diodes in their

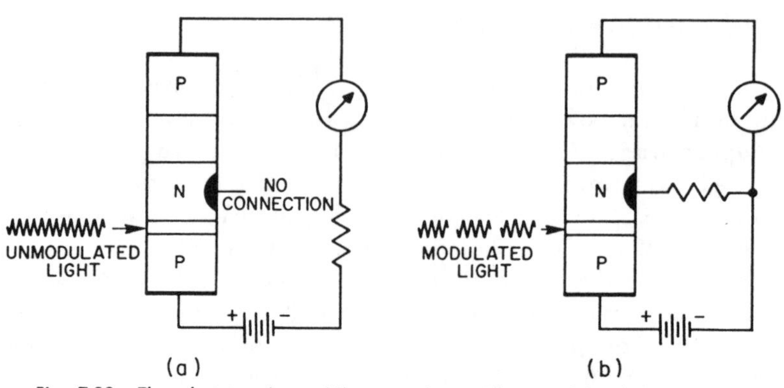

Fig. 7-29. The phototransistor: (A) connections with unmodulated light source; (B) connections with modulated light source.

reverse conduction region.) Figure 7-31 illustrates the negative resistance region. Negative resistance is a circuit parameter which enables amplification and oscillation to be obtained. This is so because power is *supplied* to an appropriate circuit rather than dissipated as with ordinary, or "positive" resistance. Indeed, it can be shown mathematically that negative resistance is present as a circuit parameter in all amplifiers and oscillators.

The tunnel diode consumes on the order of one-hundredth to one-thousandth of the d-c power required by transistors. This device operates over a temperature range much greater than can be obtained from either germanium or silicon transistors. Tunnel diodes have been made which operate at the temperature of liquid

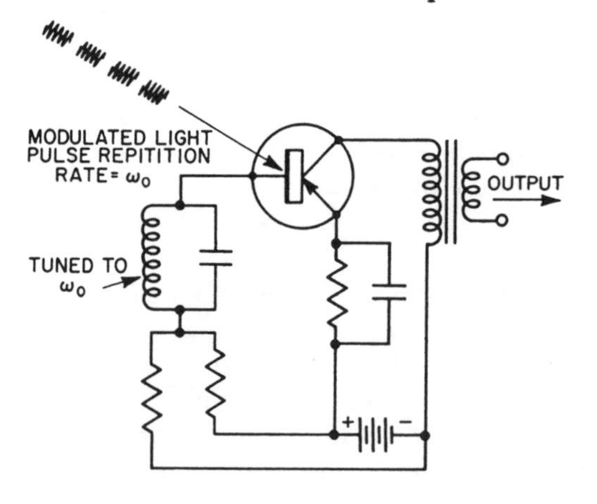

Fig. 7-30. Phototransistor circuit for high signal-to-noise ratio.

helium. Silicon tunnel diodes have shown operational capability as high as 650°F. Tunnel diodes made of other materials will permit operation at even higher temperatures. In addition to silicon and germanium, many other semiconductors appear suitable for fabrication of this amazing device. Some which have been experimented with are indium antimonide, gallium antimonide, gallium, and gallium arsenide.

The tunnel diode is very much less susceptible to damage from nuclear radiation than is the transistor. Units of 2000-mc capability are readily made, and in principle it appears feasible that the frequency response may be extended to 10 mc, and even higher microwave frequencies. Amplification is produced with much less self-generated noise than in the transistor. Finally, there is every indication that the tunnel diode will ultimately lend itself well to mass production at low cost.

The tunnel diode is basically a p-n junction diode. However, the depletion layer is much thinner than prevails in ordinary

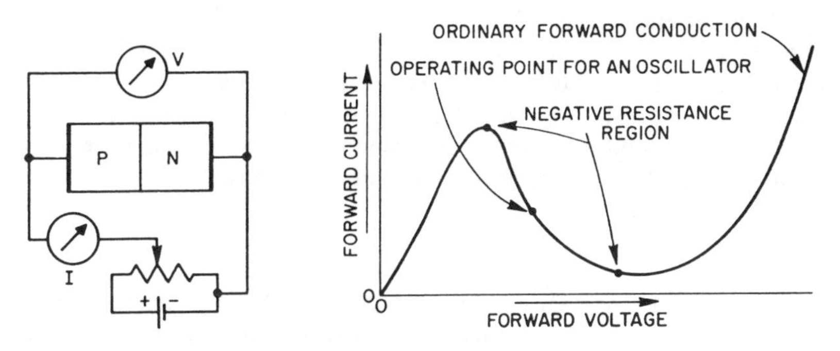

Fig. 7-31. Forward bias characteristics of the tunnel diode.

diodes or transistors, being less than 1 μin. thick. Consequently, current transfer through the junction no longer depends upon the haphazard arrival of randomly moving electrons or holes. In the tunnel diode, once an electron receives impetus to move, the effect is almost instantaneously communicated to the oppositely doped region. Specifically, an electron in the n region can cross the depletion layer at nearly the speed of light. The arrival of the electron in the p region then constitutes the flow of current through the diode. Such conduction is quite analagous to ordinary electronic conduction in metals. In a metallic conductor, we do not have to wait for the same electrons injected at one end to emerge from the opposite end. Rather, the electric effect is propagated at near light velocity by numerous "billiard ball" impacts of millions of electrons. Each electron moves a short distance, but extremely rapidly.

The movement of any single electron is similar to that occurring within the thin depletion layer of the tunnel diode. The metallic conductor will not amplify because it does not display the property of negative resistance. Negative resistance exists throughout a voltage current range in the tunnel diode by virtue of the fact that an increase in applied voltage produces a decrease in current flow through the device. This is a dynamic characteristic, only a change in voltage can receive amplification. The steady d-c bias voltage and current necessary to project operation in the middle of the negative resistance region represents expenditure of power from the d-c supply.

The basic circuit arrangement for producing oscillation in a resonant L-C tank is shown in Fig. 7-32. The negative resistance of the tunnel diode neutralizes the effect of the inherent dissipative losses of the tank circuit. When this is doen, thermal noise, or the electrical disturbance created in energizing the circuit initiates the build-up of oscillation in the resonant tank. The oscillatory amplitude increases until the peaks of the a-c wave approach the end regions of the negative resistance characteristic, whereupon equilibrium is attained. When amplification, rather than oscillation is desired, the bias is adjusted so that the numerical value of the negative resistance is somewhat less than the effective positive resistance of the circuit in which amplification is to occur. For example, in Fig. 7-32, the L-C tank could represent the primary or secondary of an i-f transformer.

THE LED

The p-n junction is the common denominator of a large number of useful semiconductor devices. We have already encountered the mechanism of charge injection into the region, and have observed that the recombination of unlike charges tends to occur as an undesirable loss. Thus, the rectification efficiency of p-n diodes and the current gain of junction transistors are less than they might be if such recombination did not take place. We know, however, that there are many instances where undesirable phenomena in one device becomes the salient operating feature of another device. This is indeed true when we consider *light emitting diodes* (LEDs) and *injection lasers*.

There is more than one way in which electrons and holes can recombine, although the net result is invariably the cancellation or annihilation of the charges. Among other classifications, such recombination can be said to be either radiative or non-radiative. As these terms imply, recombination can either emit photons of light energy or not. In the development of the

LED, it was obviously necessary to focus on conditions promoting the radiative type of electron-hole recombination. Of course, other factors are also important, such as controlling the frequency of the emitted photons, this being tantamount to specifying the operating color of the LED.

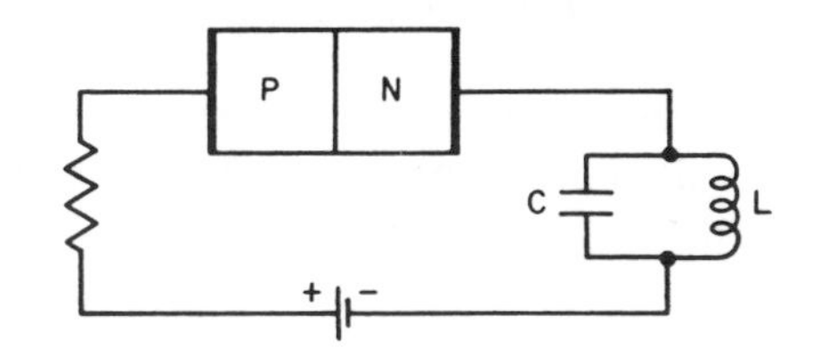

Fig. 7-32. Basic tunnel diode oscillator circuit. Oscillation occurs when the numerical resistance presented to the tank circuit equals or exceeds $\sqrt{L/C}$, the equivalent resistance resulting from parallel resonance.

It turns out that the light color generated by the LED is inversely proportional to the *band-gap voltage* of the semiconductor material. We have noted that the germanium used in diodes and transistors has a band-gap voltage of about 0.2 volt, whereas that of silicon junction devices is approximately 0.65 volt. The band-gap voltage represents the difference in energy levels between the valence band and the conduction band of the semiconductor material. The alluded band-gap voltages represent energies of infrared light. Therefore, neither the germanium or the silicon semiconductor material used in ordinary diodes and transistors would be suitable for producing visible light.

In order to make use of radiative recombination that emits visible light, it is necessary to utilize semiconductor material with higher band-gap voltages. For example, semiconductor material which will generate red light must have a band-gap of about 2 volts. For blue light, a band-gap of about 2.9 volts is required. The wavelength of red light is in the 630 to 700 nanometer region, while that of blue is about 450 nanometers. The other visible colors fall between these two extremes. (Infra-red radiation generally encompasses the 900 to 1600 nanometer portion of the spectrum.) The basic p-n diode nature of the LED is depicted in Fig. 7-33.

Because of the relationship between the band-gap voltage and the frequency of the emitted light from radiative combinations, research and development projects dealing with LEDs have been referred to as "band-gap engineering." The practical ramification of this term is that new compounds and alloys have had to be investigated. For the most part, these have turned out to be III-V semiconductors, i.e. materials compounded from elements in group three and in group five of the periodic table. And, quite often, the optimum material comprises three or more chemically-bonded elements. Thus, we find such LED semiconductor materials as gallium arsenide phosphide, which yields orange light, gallium phosphide (GaP) for green, and silicon carbide (SiC) which produces a rather anemic blue light. After many years of frustrating effort, a much better semiconductor material was found in the form of zinc-doped indium gallium nitride (InGaN) and aluminum gallium nitride (AlGaN). This complex material produces about one hundred times the luminous intensity of the blue light that could be had from the earlier silicon-carbide LED.

It is interesting to contemplate that the process responsible for LED performance is essentially the inverse of that occurring in the solid-state photocell. In the photocell, light photons supply the energy needed to produce electron-hole pairs, which then become the delivered electric current.

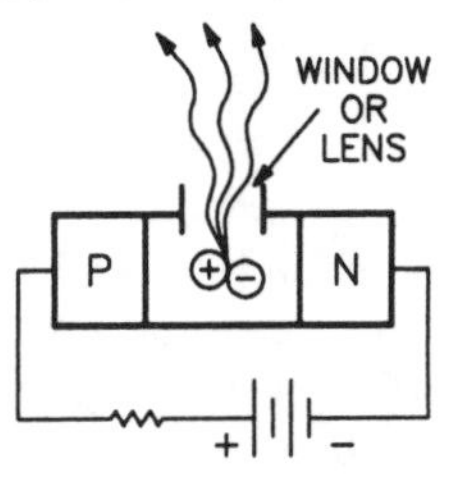

Fig. 7-33. The LED is basically a forward-biased p-n diode. However, the LED is fabricated from appropriate semiconductor material to promote radiative recombination of charge carriers and to deliver the resultant photonic energy (light) into the environment. Radiative recombination occurs in ordinary diodes, but represents a dissipative loss and the photons do not emerge from the device as light.

Conversely, in the LED, the recombination of electrons and holes which have been injected in the depletion region give up their energy in the form of emitted light. Both of these p-n devices can function as rectifying diodes even though rectification is not the normally sought performance parameter.

The efficiency of LEDs have been so greatly improved that they now compete with incandescent lamps for certain applications. Thus, it has been found that red LEDs can outshine red-filtered incandescent lamps in automotive brake and tail lights. Such bright LEDs also qualify for use in traffic lights and in advertising displays. Unlike incandescent lamps, LEDs do not have catastrophic failure modes. Rather, it is claimed that a million hours of continuous operation would just suffice to reduce brightness by half.

THE SEMICONDUCTOR INJECTION LASER

The production of coherent beams of light energy is certainly one of the notable achievements of our technological era. Whereas ordinary sources of light emit sporadically timed groups of random frequencies, ideal coherent radiation comprises a single frequency where all radiative processes are synchronized at the atomic level. The resultant light production is concentrated in a narrow beam instead of being scattered in time and space. This is tantamount to saying that the power density in the coherent beam is exceedingly high. Indeed, such beams can quickly vaporize holes in diamonds, metals, and ceramics. Because of the very low divergence of the beam, a relatively small circular area can be illuminated on the surface of the moon. Lasers, which generate these coherent light beams, have a myriad of applications in industry, communications, and in medicine. It is not only feasible to weld large structural members, but a detached retina in the eye can also be "welded" back in place.

The term *laser* is an acronym standing for "light amplification by stimulated emission of radiation." Although there are a bewildering number of gaseous, liquid, and solid lasers, most operate on the same basic principle. A laser is essentially a tightly compacted electro-optical system in which artificially triggered fluorescence is involved in a feedback process for stimulating further fluorescence. Because of this positive feedback, the emerging beam of light, though of a small cross-sectional area, represents inordinately high power density. It is as though the simultaneously occurring photons were tightly compressed about a unidirectional line of flight.

Laser action comes about in almost all lasers by inverting the populations of electronic energy levels in atoms, as illustrated in Fig. 7-34. This may sound esoteric, but actually is easy to do. A common way of accomplishing this is to expose the lasing element, which could be a ruby crystal with a small percentage of impurity atoms, to a powerful light source. The impurity atoms absorb some of the light and their orbital electrons are thereby excited to higher energy states than they normally occupy. However, some of these excited states are not stable and the electrons tend to fall back to their normal energy levels. In so doing, they give up their temporarily acquired extra energy as radiated photons. This mechanism accounts for the phenomena of *fluorescence*. A simple demonstration of fluorescence can be had by observing the beautiful colors produced when various minerals are illuminate with ultraviolet light.

Although fluorescence is the first step in attaining operation in the lasing mode, it, though necessary, is not sufficient. The next step is to cause self-excitation of the fluorescent process in order to greatly build up the optical power level. Although the practical means of doing this was long in coming, it is clear in retrospect that nature was on our side. This is because the restimulated atoms are especially responsive to the very level of photon energy they release when fluorescing.

In order to bridge the gap between the fluorescing and lasing modes of operation, a simple optical trick is used. The material undergoing fluorescence is enclosed in an imperfect optical cavity. The optical cavity is imperfect in the sense that at least one of the two mirrors at its ends are purposely made "leaky" so that a portion of the light energy can penetrate into the environment, thereby becoming the usable laser beam. As the light within the cavity bounces from mirror to mirror, it restimulates fluorescence during each pass. The power level of the self-excited fluorescence becomes very high from this process and is finally limited because of various dissipative losses, as well as by leakage of part of the energy through the imperfect mirror. A significant aspect of this process is that the restimulated atoms are all excited in phase. This accounts for the coherent nature of the emergent laser beam.

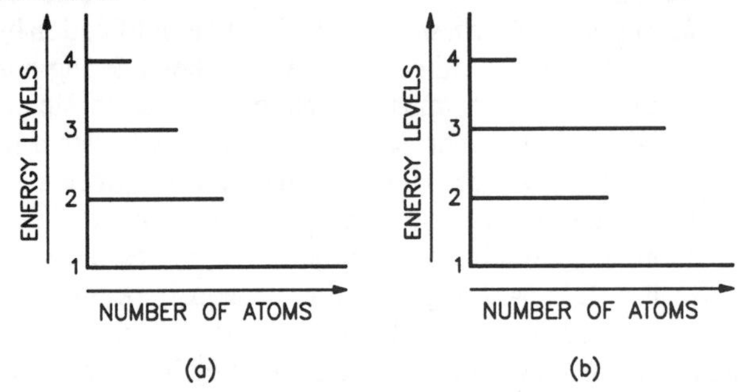

Fig. 7-34. The concept of population inversion is a basic phenomenon of lasers. (A) Normal levels of the atoms in a substance. (B) Population inversion of energy levels when the atoms are excited from absorbed energy. Situation (B) tends to be unstable and the excited atoms revert to their normal distribution of energy levels. This comes about as the electrons of the atoms shed their stimulated energies. In so doing, they radiate photons (light energy).

Once one grasps these basic ideas about lasing action, it should come as no surprise that the LED can be made to lase via a few modifications in its structure. This is because the generation of the light beam occurs from energy transformations very similar to fluorescence. The light photons emitted by radiative recombination in the p-n region of the semiconductor have frequencies (colors) governed by the energy levels (band-gaps) of the recombining charges. Thus, the luminescence produced resembles fluorescence except that an external light source is not needed. Instead of the external light source, we simply inject minority charges into the p-n region by forward-biasing the LED diode.

Next, we must deal with the phenomenon of *population inversions*. It fortunately happens that population inversions of energy levels does indeed take place if the LED bias-current is high enough. Of course, in the LED, no practical use is make of population inversion in the form of lasing. Having covered this much ground, it seems reasonable to anticipate the semiconductor laser by associating a leaky mirror optical cavity with the LED. This is essentially what has been done. In operation, the so-called *semiconductor injection laser* will operate as an ordinary LED at low forward-bias current. The emitted light is then noncoherent. If we increase the bias current, a point will be reached at which the lasing mode abruptly commences and the emergent beam is coherent in nature. It goes without saying that the reduction of the aforementioned principles to practical hardware has been a tremendous accomplishment in light of the precision required from tiny structures, and the ability to accurately manipulate the chemical composition of the active regions.

It was almost too good to be true that the semiconductor injection laser came on the scene in time to be incorporated in the evolving fiber-optic communications system. Because of its compact dimensions and its reliability, the semiconductor injection laser was just what the doctor ordered. Additionally, it was feasible to design the lasers to work in the infra-red region of 1300 to 1600 nanometers. Within this spectral region, the optic fibers could be engineered for minimum attenuation and/or minimum chromatic dispersion. Translated into the operational features of such a communications system, this implies both transmission over longer distances and higher speed performance.

A nice feature of the semiconductor injection laser is the ease at which high-rate digital modulation can be accomplished. It is merely a matter of turning the device current on and off. Signaling rates in the vicinity of 500 megabits per second can be imparted in this manner. Moreover, the injection laser together with the modulator devices can both be implemented in a single integrated circuit. Success has been thus realized with both FET and bipolar transistors. Such a system can operate from a 5 volt supply, dispensing with the high voltage, large capacitors, and the massive optical pump hardware needed for other types of lasers. The simplicity of digitally modulating the semiconductor injection laser is depicted in Fig. 7-35. For the sake of speed, the modulator could, in turn, be driven from an emitter-coupled logic source of signaling pulses.

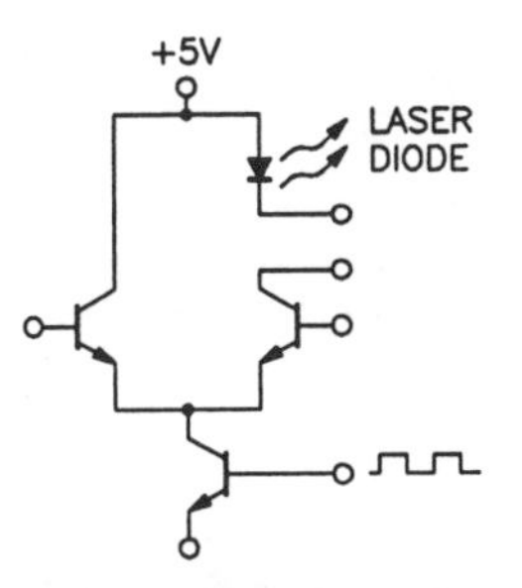

Fig. 7-35. Modulation scheme for semiconductor laser diode. Rather than being comprised of discrete devices, this scheme represents a monolithically integrated circuit. Similar ICs have been made with FET control devices. Optimum performance readily results from a 5 volt DC supply.

THE THERMOELECTRIC COOLER

Refrigeration can be produced by passing current through a bismuth telluride to metal junction. The basic phenomena has been known for a long time, and is the well-described *Peltier Effect* one finds in physics texts. However, prior to the semiconductor era, the alluded junctions were comprised of two different metals. These thermocouple devices are useful for many instrumentation functions because they develop an EMF in response to impinging heat. It was always true that the inverse effect also existed, that is, a cooling effect could be produced by passage of a current through the junction. However, the cooling effect was not very pronounced and tended to be masked by an actual temperature rise from dissipative losses.

One should be aware that the use of the term junction pertaining to these devices does not signify the p-n junction we tend to associate with semiconductor devices. In this instance, junction simply implies a physical interface. Of importance is the fact that the substitution of appropriate semiconductor material for one of the metals previously employed, greatly enhances the Peltier Effect. The practical result is that passing current one way through such a junction can appreciably cool one side of the junction and any object in physical proximity to it. At the same time, the other side of the device will experience a rise in temperature and is ordinarily equipped with a heat sink. Reversing the current surprisingly alternates the cooling and heating plates. An example of a semiconductor thermoelectric cooler is shown in Fig. 7-36.

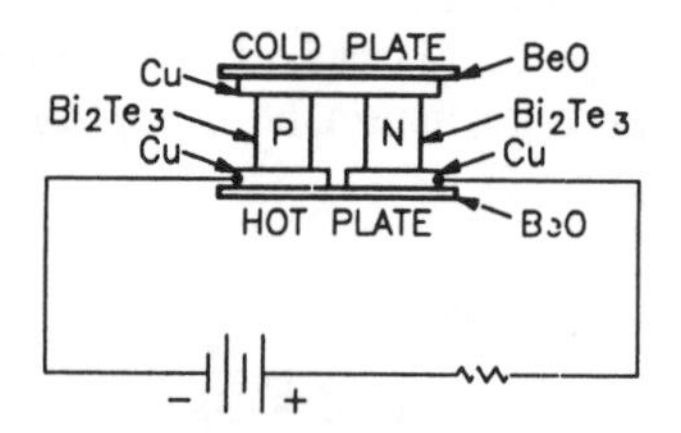

Fig. 7-36. Basic arrangement of the thermoelectric cooler. The p-type and n-type bismuth telluride (Bi_2Te_3) semiconductor material is in physical contact with the copper (Cu) plates, but these contacts are not p-n junctions. The thin beryllium oxide (BeO) sheets combine electrical insulation with high thermal conductivity. Actual physical dimensions of a thermoelectric cooler could be a 1-1/2" square about 1/8" thick.

What has actually been described is a heat pump. Heat energy is removed from one plate of the device and delivered to the opposite plate. The fact that the process can be reversed by reversing the polarity of the current allows servo control of temperature. This is very useful for temperature-stabalized ovens, crystal oscillators, voltage references, and various scientific applications. Keep in mind that the electrical interface of these so-called junctions is neither a p-n junction, nor is it a Schottky barrier which it resembles because of the semiconductor/metal contact. Rather, it is more nearly just a ohmic connection similar to the terminal connections in ordinary diodes and transistors. This is revealed by the two-way passage of current the device exhibits. Heavy doping of the semiconductor material accounts for these ohmic contacts.

OTHER NON-JUNCTION SEMICONDUCTOR DEVICES

Devices that respond to magnetic fields are both intriguing and very useful. The *Hall Effect* was discovered in metallic conductors exposed to a magnetic field, but later it was found that the effect is much more pronounced in certain semiconductors, particularly in those having high electron mobility. Note in Fig. 7-37*A* that the current sent through the slab of material, the magnetic field, and the sensing probes of the output circuit are each mutually perpendicular to one another. Because the probe that senses the output of the device connects to the midpoints of the material, one might not expect to detect the longitudinal current provided by the d-c source. One might look at the setup as a balanced bridge; even though a heavy current flows lengthwise through the material, it would seem that no portion of it would be indicated by the output voltmeter.

Without the magnetic field, such a null condition would indeed exist. The situation however, is altered by the magnetic field. A motor force is exerted on the electron flow just as if these electrons were a thin current-carrying wire. This being the case, a denser electron region will exist near one of the sensing probes than in the vicinity of the other. This difference in electron density will then manifest itself as a potential difference, and the output voltmeter will register a reading. Further investigation would show that the voltage developed in this way is directly proportional to both the strength of the longitudinal current and the strength of the magnetic field. Moreover, the polarity of the detected voltage will reverse if either the longitudinal current or the direction of the magnetic field is reversed.

An even simpler sensor for magnetic fields is shown in Fig. 7-37*B*. The conductivity of certain bismuth alloys directly responds to magnetic field strength. However, in this case, the direction of the magnetic field is of no consequence. The shortcoming of this type of response is the large temperature dependency of the resistance in the materials which respond strongly to magnetic fields.

Because of the responses revealed by devices such as that depicted in Fig. 7-37*AB*, it would be only natural to imagine transistors which would operate from an input signal represented by a magnetic field. Such "magistors" have appeared on the market, but were considered inferior in performance to other devices and never gained popularity in practice. It is not inconceivable, however, that a vastly improved magnetically actuated semiconductor device might again be marketed. (It is also interesting to note that magnetically actuated vacuum tubes once made a short-lived appearance.)

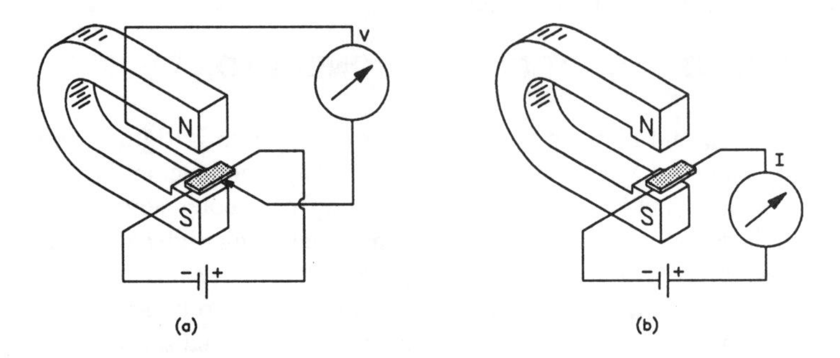

Fig. 7-37. Additional non-p-n junction semiconductor devices. In devices of this nature, it is the high charge mobility of the semiconductor material that is responsible for a good response. (A) Hall Effect device. The same action that moves the armature conductors in a DC motor displaces the electrons toward one edge of the semiconductor material. This, in turn, manifests itself as a voltage difference across the face of the material. (B) Simple device in which resistance is a function of the magnetic field strength.

THE SCHOTTKY DIODE

The Schottky diode displays rectification characteristics, but this behavior does not stem from a p-n junction. One might say that this device incorporates a half p-n junction, but such a description must not be pursued too literally. Most Schottky diodes have been fabricated from N-type silicon in contact with a metallic plate, although germanium and gallium arsenide types have at least at one time also been available. The interface between the semiconductor and the metal is also known as a "hot-carrier junction." The basic mechanism of rectification is not dissimilar to that responsible for the demodulation of AM signals by the crystal detectors of early radio sets. A practical difference is that the crystal detector was a point-contact device, whereas the Schottky diode is an area-contact device and is also able to handle large currents. The salient operating feature of the Schottky diode is that it is an efficient rectifier for the higher frequencies. (In power supply practice, this tends to be anything above about 20 kilohertz.) Schottky's are RF starts. Another much sought-after feature of the Schottky diode is its relatively low forward-voltage drop. This too, enhances its rectification efficiency.

Conventional p-n rectifying diodes perform well enough below 20 or 30 kilohertz, but then yield progressively degraded performance with increasing frequency. During the forward-conduction portion of an alternating current cycle, the p-n diode stores minority charges in its p-n (depletion) region. These charges require time to deplete. Unfortunately they manifest themselves as forward-conduction current during the onset of the reversal of the applied a-c voltage, at the very time when rectification depends on negligible conduction. At low frequencies, this phenomenon has negligible consequences because there is lot of time for recovery from the stored charges. But, at higher frequencies, this recovery period then constitutes an appreciable portion of the a-c cycle. This seriously impairs rectification efficiency, causes temperature rise in the diode, and also generates EMI and RFI because of the snap-action of the recovery characteristic. Moreover, the poorer rectification that ensues can play havoc with the electrolytic filter capacitors. The charge storage effect is shown in Fig. 7-38.

The Schottky diode, on the other hand, is a majority carrier device. There are not minority carriers to be stored in the material. As one may suspect, there are some trade-offs in designing with Schottky devices. They are relatively low voltage devices because their reverse currents increase rapidly with applied voltage. By the same token, heat dissipation must be well-managed as the reverse current will also go up rapidly with temperature. Much progress has been made in extending performance perimeters, however. Early units were specified for 10 or 15 volts, and limited to very moderate temperature rises. More recent diodes can have current ratings of tens or hundreds of amperes, operate at 50 volts, and sustain "junction" temperatures at least 125° C.

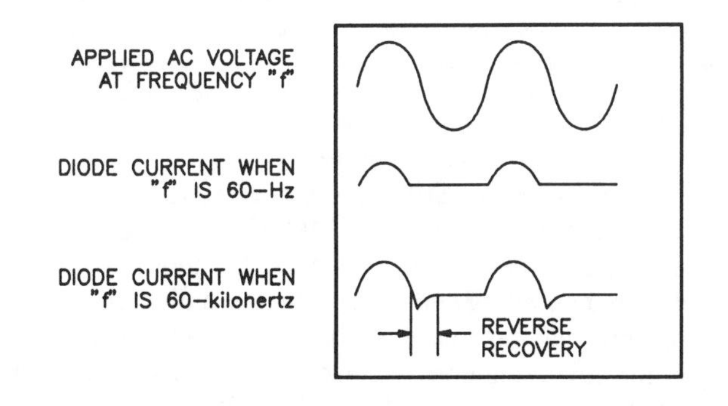

Fig. 7-38. The stored charge effect in p-n junction diodes. Rectification is degraded because conduction occurs for part of the reverse time of the applied AC voltage.

THE POWER MOSFET

One of the most significant developments in semiconductor technology since the conception and evolution of the bipolar junction transistor has been the *power MOSFET*. This type of transistor is unique in that its basic operation is not dependent upon p-n junctions. It traces its heritage to the previously described unipolar field transistor, later to become the junction field effect transistor (JFET). *MOSFETs* and power MOSFETs continue the use of an electric field to modulate the conductivity of a channel, but dispense with the p-n junction for the gate. Rather, a metal or other conductive plate is separated from the channel by a thin insulating layer of oxide. In other words, the gate section of the device is a capacitor. Indeed, the input impedance of MOSFETs behave just as one would expect from a capacitor. This endows the MOSFET devices with extremely high input impedance at low frequencies and at DC. That is, there is virtually no current consumption from the signal source until one works in the high kilohertz and in the megahertz region. Another way of saying this is to state that MOSFETs exhibit tube-like performance. The structural aspects of the MOSFET are illustrated in the simplified drawing of Fig. 7-39.

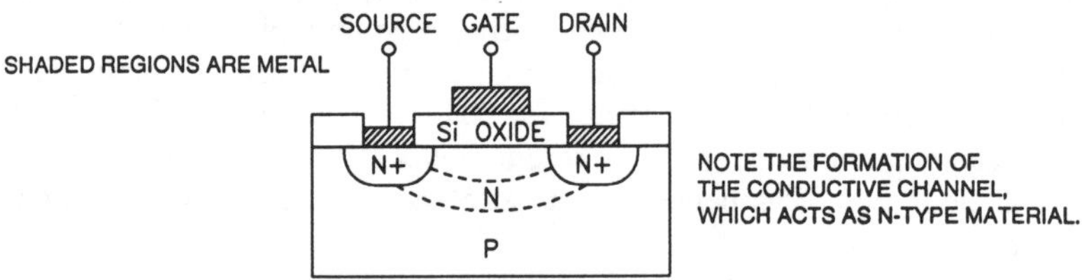

Fig. 7-39. Simplified sketch of the basic MOSFET device. Although there are interfaces between p and n regions, basic transistor action is not dependent upon p-n junctions. The gate input section is a capacitor. The output section between source and drain behaves as a channel in which conductivity is controlled by the electric field from the gate. This is because the passively-p material in the channel is effectively converted to n-type during operation. Accordingly, the output section simulates the behavior of a simple n-type silicon bar with no p-n junctions.

There has been much technical literature devoted to comparisons between the *bipolar transistor* and the power MOSFETs. Such write-ups invariably become outdated in a short time because the industry manages to close the gap between the leading and trailing device, and often reverses the relative merits of some features in the two devices. Thus, there has been a ongoing competition involving cost, voltage, current, frequency capabilities, and reliability between the two technologies. Nonetheless, the following generalities and tendencies are in order because of the different fabrication in these rival semiconductor devices:

1. The bipolar transistor is vulnerable to such failure modes as thermal runaway and secondary breakdown. The power MOSFET is relatively

immune to such catastrophic destruction. One reason is that its basic operation involves no p-n junctions. Another reason is that temperature rise in the MOSFET is accompanied by increased resistance of the channel, thereby partially limiting the effect of over-current. The opposite temperature coefficient works in the bipolar transistor to cause thermal runaway.

2. The power MOSFET is a majority carrier device so it is not plagued by minority carrier storage. This tends to make it a better candidate for high-frequency switching applications. In general, high frequency capability is less costly and involves fewer trade-offs in the power MOSFET than in a bipolar transistor. Just the same, a wise designer will check the state of the art before choosing between the two types.

3. The power MOSFET is vulnerable to gate destruction from excessive input voltage or from transients riding on the gate voltage. Such destruction is analogous to dielectric rupture in capacitors. Gate destruction can also result from the electro-static charge picked up in one's body and transferred to the MOSFET during installation or from handling. With appropriate precautions however, this danger can be greatly minimized.

4. At higher voltage ratings, say in excess of several hundred, the output voltage drop in the power MOSFET may be high enough to seriously impair operating efficiency. By contrast, the collector voltage drop in all bipolar transistors remains relatively low.

5. Power MOSFETs are easy to operate in parallel without need for ballast resistances. Current-sharing tends to take place automatically because of the positive coefficient of resistance with respect to increasing temperature.

6. The power MOSFET, like the bipolar transistor, is available in "n" and "p" complimentary types.

7. The predominant semiconductor material for power MOSFETs is silicon. However, for UHF and microwaves, gallium arsenide provides inordinately low noise-temperatures making such devices very useful for the reception of satellite transmissions.

8. Most power MOSFETs are enhancement types, i.e. without threshold gate-bias, the device is off. However, depletion types are also available; these are normally on, like many vacuum tubes. No counterpart exists with bipolar transistors.

9. The gate threshold voltage of enhancement type power MOSFETs has been progressively reduced so that 5 volt operation of certain units is easily implemented.

10. Special power devices known as IGBTs (insulated gate bipolar transistors) have become available in which the output voltage drop compares favorably with that of high-voltage bipolar transistors. The trade-off is that the IGBT has limited frequency capability, on the order of 50 kilohertz. The

output section of these devices, unlike conventional MOSFETs, contains a p-n junction. With MOSFET input and bipolar output, the IGBT is a hybrid device with salient features of both types of transistors.

THE GUNN DIODE

The *Gunn diode* is a unique microwave oscillating device. In the first place, it is a diode only in the sense that there are two connections to it. Otherwise, it exhibits no rectifying properties. Nor does it comprise regions analogous to those of either thermioni or solid-state diodes. Superficial inspection of its architecture reveals a simple sandwich of gallium arsenide layers; however, the outer two are heavily doped so that they act as ohmic connections to the inner layer. Indeed, the action takes place only within the inner layer. One must not look for any resemblance to either a p-n junction or a Schottky junction. The discovery of its characteristics came about accidentally. The physics underlying its behavior was formulated later. An everyday acquaintance with the device is useful, despite its simplicity, to recognize it as bulk semiconductor material that exhibits negative resistance. The basic structure of the Gunn diode is shown in Fig. 7-40.

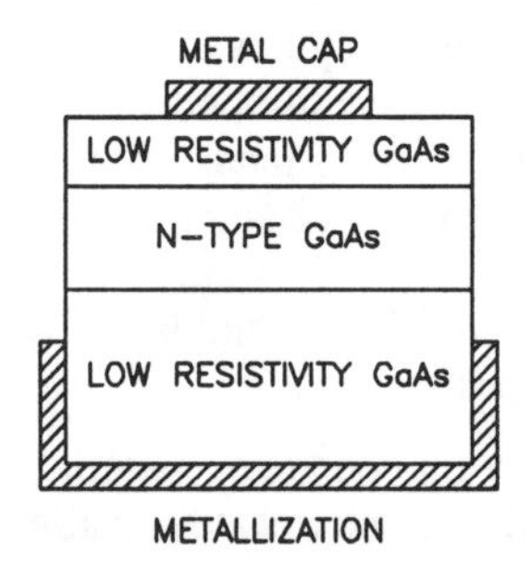

Fig. 7-40. Basic structure of the Gunn diode. This sandwich involves no p-n junctions. The two outer gallium arsenide (GaAs) regions are heavily doped and provide ordinary ohmic connection to the inner GaAs layer, as well as to the external bias circuit. The active region of the device develops within the inner layer as bias is applied and oscillation ensues.

For steady-state oscillation, the Gunn diode finds many applications in radar, communications, and instrumentation. At the same time, it is very inefficient for such continuous duty operation; four or five times more d-c power must be dissipated as heat than is available as microwave output. Incidentally, the Gunn diode has been worked from several GHz to over 100 GHz. As might be expected, the device is fabricated with a self-contained heat-sink. Power outputs for continuous oscillation generally range from 5 milliwatts to about 1 watt. Reversing the polarity of the applied d-c bias voltage will likely result in catastrophic destruction. This might appear to contradict the previous allusion to electrical symmetry of the device. The bulk semiconductor material should not care which way current goes through it. This is true enough, but from a physical and thermal standpoint, the device is far from being a symmetrical structure as one ohmic region is very thin and the other is relatively thick. Reversing current flow does not blow out any junction, but brings about quick thermal failure because the heat-sink is then too far away from the active region where the temperature rise of the device is being developed.

The active layer within the bulk material displays a thickness inversely proportional to the generated frequency. This strange feature is suggestive of piezoelectric phenomena. However, it turns out that an entirely different manifestation of physics is responsible for this unexpected relationship. The logarithmic graph of Fig. 7-41 depicts the relationship between thickness or width of the active layer and the frequency. The variation of bias-voltage with frequency is also shown. It is evident that the d-c voltage source for the bias should be closely regulated.

Once it is accepted that bulk semiconductor material can behave as a negative resistance, it is easy enough to see how the Gunn diode can be used as a microwave oscillator. Oscillation theory postulates that any oscillator, regardless of type, can be represented by an equivalent arrangement of a negative resistance associated with a resonant tank circuit (or with a resonant cavity). The ordinary explanation is that the negative resistance cancels the positive or dissipative resistance, and the resonant tank then has freedom to build up and sustain oscillatory action. The initiation of the build up process can originate from noise transients, however, it is only natural to ponder why the negative resistance should assert itself in the first place.

Gallium arsenide, indium phosphide, and a few other materials appear to have two pronounced conductive bands, as opposed to silicon and germanium, which for practical purposes, show evidence of just one. Recall in our discussion of LEDs and semiconductor injection lasers that we dealt with the concept of the energy levels of the orbital electrons in the atom. When electrons in the balance band absorb energy, some of them are raised into the conduction band. If this simple process is underway in gallium arsenide or indium phosphide and still more energy is added (say from higher voltage applied to the material) increasing numbers of electrons become elevated to higher energy levels and then begin to occupy the second conduction band. However, the mobility of electrons in the second conduction band is less than in the first conduction band. Lower electron mobility in a semiconductor implies less current conduction. That is, the effective resistance increases. Overall, the result of further increasing the voltage is to cause the current to go down instead of up. It is just such behavior which is said to stem from negative resistance.

This cause and effect relationship is basic to Gunn diode behavior. It does not account for all observed features however. For example, the Gunn diode, when operating in the Gunn mode, can generate microwave energy of nearly predictable frequency when working into a resistive load — the ordinarily required resonant tank can be dispensed with. (In practice, a resonant line or cavity is invariably used in order to achieve better efficiency, greater wave purity, and to impart at least a small tuning range.)

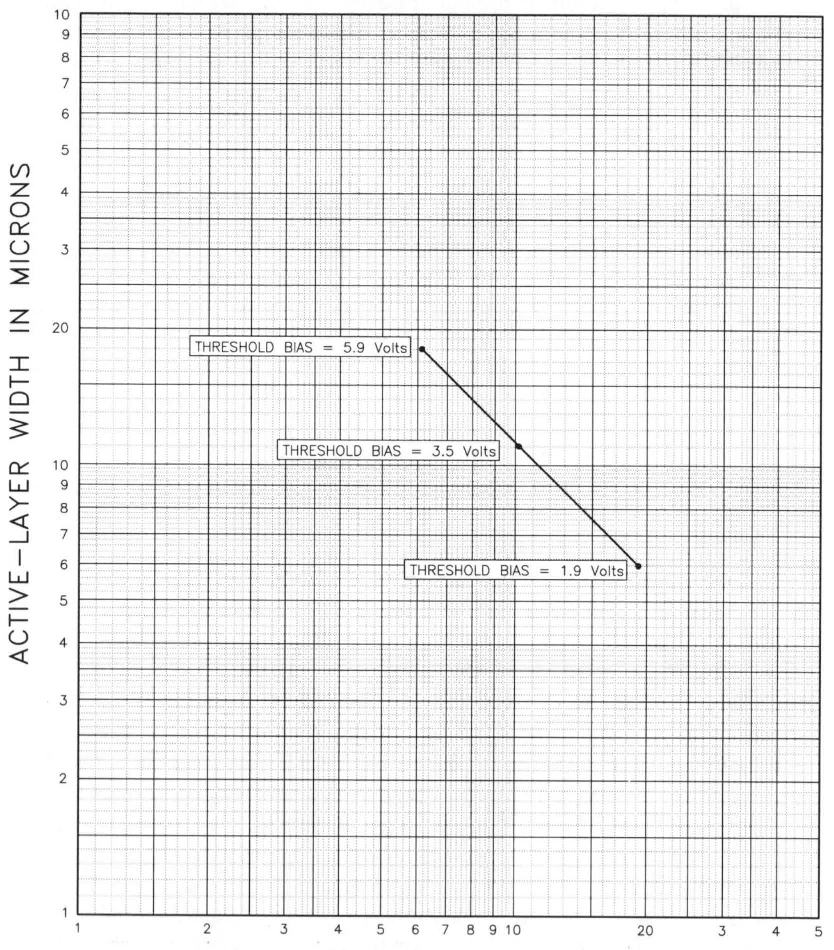

Fig. 7-41. Width of active layer in Gunn diode as a function of the frequency. Note also the inverse relationship between the frequency and the threshold bias voltage.

For steady-state oscillation, the Gunn diode finds many applications in radar, communications, and instrumentation. At the same time, it is very inefficient for such continuous duty operation; four or five times more d-c power must be dissipated as heat than is available as microwave output. Incidentally, the Gunn diode has been worked from several GHz to over 100 GHz. As might be expected, the device is fabricated with a self-contained heat-sink. Power outputs for continuous oscillation generally range from 5 milliwatts to about 1 watt. Reversing the polarity of the applied d-c bias voltage will likely result in catastrophic destruction. This might appear to contradict the previous allusion to electrical symmetry of the device. The bulk semiconductor material should not care which way current goes through it. This is true enough, but from a physical and thermal standpoint, the device is far from being a symmetrical structure as one ohmic region is very thin and the other is relatively thick. Reversing current flow does not blow out any junction, but brings about quick thermal failure because the heat-sink is then too far away from the active region where the temperature rise of the device is being developed.

The active layer within the bulk material displays a thickness inversely proportional to the generated frequency. This strange feature is suggestive of piezoelectric phenomena. However, it turns out that an entirely different manifestation of physics is responsible for this unexpected relationship. The logarithmic graph of Fig. 7-41 depicts the relationship between thickness or width of the active layer and the frequency. The variation of bias-voltage with frequency is also shown. It is evident that the d-c voltage source for the bias should be closely regulated.

Once it is accepted that bulk semiconductor material can behave as a negative resistance, it is easy enough to see how the Gunn diode can be used as a microwave oscillator. Oscillation theory postulates that any oscillator, regardless of type, can be represented by an equivalent arrangement of a negative resistance associated with a resonant tank circuit (or with a resonant cavity). The ordinary explanation is that the negative resistance cancels the positive or dissipative resistance, and the resonant tank then has freedom to build up and sustain oscillatory action. The initiation of the build up process can originate from noise transients, however, it is only natural to ponder why the negative resistance should assert itself in the first place.

Although the negative resistance characteristic is one of the salient features of the Gunn diode, the transit time of domains of electrons actually produce the oscillations within the active layer of the device. In this respect, it differs from other negative resistance devices, such as the tunnel diode. The Gunn diode is accordingly known as a TEO (transferred electron oscillator). The oscillatory process already described is known as the transit time, or Gunn mode. It is usually the operational mode for low power continuous-wave operation.

The Gunn diode can also be operated in the delayed transit time mode. This mode is used for low duty-cycle pulsed operation and enables high peak powers to be developed. With pulse duty-cycles on the order of one percent, up to several hundred watts of peak power can be obtained. With an appropriate combination of bias voltage and RF amplitude, the continuity of the moving electrons domains within the active layer is interrupted. Shock-excitation of the resonant tank then becomes the dominant timing impetus of oscillation. (We are reminded of the pulse-type excitation of the tank circuit in a class C amplifier.) One of the practical manifestations of this operational mode is that oscillation frequency becomes largely governed by the tank resonance. Thus, such a pulsed oscillator can be tuned over an octave frequency range.

Gunn diodes should be specified for their intended mode of operation (CW or pulsed). Also, the frequency or frequency range, as well as the power level must be indicated when ordering. One should be aware that the devices are made with anode heat-sinks as well as with cathode heat-sinks. This consideration affects the optimum type of mounting within the waveguide or coaxial cavity. Once properly implemented and operated, the Gunn diode is an elegantly simple device of high reliability. However, during setup and experimental procedures, this device is too costly to be treated as "just another diode." Any carelessness with regard to polarity reversal, excessive current, or inadequate thermal conditions is not likely to be rewarded with a second chance.

index

D

E

F

G

H

I

☛ Dear Reader: *We'd like your views on the books we publish.*

PROMPT® Publications, an imprint of Howard W. Sams & Company, is dedicated to bringing you timely and authoritative documentation and information you can use.

You can help us in our continuing effort to meet your information needs. Please take a few moments to answer the questions below. Your answers will help us serve you better in the future.

1. What is the title of the book you purchased? ____________
2. Where do you usually buy books? ____________
3. Where did you buy this book? ____________
4. Was the information useful? ____________
5. What did you like most about the book? ____________
6. What did you like least? ____________
7. Is there any other information you'd like included? ____________
8. In what subject areas would you like us to publish more books?

   (Please check the boxes next to your fields of interest.)

   - ❑ Amateur Radio
   - ❑ Antique Radio and TV
   - ❑ Audio Equipment Repair
   - ❑ Camcorder Repair
   - ❑ Computer Hardware
   - ❑ Computer Programming
   - ❑ Computer Software
   - ❑ Electronics Concepts Theory
   - ❑ Electronics Projects/Hobbies
   - ❑ Home Appliance Repair
   - ❑ TV Repair
   - ❑ VCR Repair

9. Are there other subjects not covered in the checklist that you'd like to see books about?

   ____

10. Comments ____________

____

____

Name ____________

Address ____________

City ____________ State/Zip ____________

Occupation ____________ Daytime Phone ____________

Thanks for helping us make our books better for all of our readers. Please drop this postage-paid card in the nearest mailbox.

For more information about PROMPT® Publications,

see your authorized Sams PHOTOFACT® distributor.

Or call 1-800-428-7267 for the name of your nearest PROMPT® Publications distributor.

Imprint of Howard W. Sams & Company

2647 Waterfront Parkway East Drive,

Indianapolis, IN 46214-2041